Chemical Graph Theory

Volume II

Author

Nenad Trinajstić

Professor of Chemistry
The Rugjer Bošković Institute
Zagreb, Croatia
Yugoslavia

CRC Press, Inc.
Boca Raton, Florida

Library of Congress Cataloging in Publication Data

Trinajstic, Nenad,1936-
  Chemical graph theory.

Includes bibliographical references and index.
    1. Chemistry—Mathematics. 2. Graph theory.
I. Title.
QD39.3.M3T74   1983   541.2'2'015115   82-12895
ISBN 0-8493-5273-8 (v.1)
ISBN 0-8493-5274-6 (v.2)

Direct all inquiries to CRC Press, Inc., 2000 Corporate Blvd., N.W., Boca Raton, Florida, 33431.

© 1983 by CRC Press, Inc.

International Standard Book Number 0-8493-5273-8 (Volume I)
International Standard Book Number 0-8493-5274-6 (Volume II)
Library of Congress Card Number 82-12895
Printed in the United States

To my brother Ivan

PREFACE

The greater part of this book was written during my visit to the Department of Chemistry, the University of South Carolina at Columbia in late autumn 1979 and winter 1980. I felt for some time that an introductory book on chemical graph theory is needed, because it is considered by many as a new branch of chemistry, though its beginnings go back to over 100 years ago. Since I was asked to deliver a series of lectures on chemical graph theory, I used the notes for these lectures as an embryo from which the present book has evolved.

This book is entitled *Chemical Graph Theory* because we will be mostly concerned with the handling of the chemical (molecular) graphs, i.e., mathematical diagrams representing molecular structures. Thus, chemical graph theory is engaged with analyses of all consequences of a connectivity in a system, e.g., bonds in molecules or reaction paths in chemical transformations. Ordinarily, graph theory does not produce numerical data (just as group theory), it uses available data and searches for regularities that can be attributed to a combinatorial and topological origin. Therefore, here is a good opportunity to indicate clearly for the chemical community at large, in order to avoid possible misunderstandings of the use and potential of (chemical) graph theory, that graph theoretical methods and proof techniques should be expected to be used as a complementary approach where the topology and the combinatorial nature play an important role, in parallel to the application of the group theory to problems where symmetry is an important characteristic of the system. Besides in the chemical graph theory we are allowed to rely on the intuitive understanding of many concepts and theorems rather than on formal mathematical proofs.

The roots of chemical graph theory may be found in the work by Higgins,[1] who has used first chemical graphs (albeit not recognized as such) for representing molecules; Kopp,[2] who studied the additive properties of molecules; and Crum Brown,[3] who has made use of quite a modern graphic notation for chemical compounds. Other early contributors to chemical graph theory are Laurent,[4] who discovered that the number of atoms with odd valency in a molecule is always even; Cayley,[5-7] who developed the mathematical theory of isomers (alkanes) using graphs called trees; Flavitzky,[8] who calculated the number of isomers of saturated alcohols; and Sylvester,[9] who first pointed out the analogy between chemistry and algebra. Sylvester's work is especially important to us because he clearly expressed a belief that there is a common ground for the interchange of mathematical and chemical ideas which may lead to new development of both fields what was the case in many instances since his time.[10-15]

This book will be mainly concerned with the structural aspects of chemical graph theory. It may be partitioned in three parts: (1) the elements of graph theory needed for the understanding of what follows; (2) the topological aspects of Hückel theory, resonance theory, and theories of aromaticity; and (3) the applications of chemical graph theory to the structure-property and structure-activity relationships and to the isomer enumeration. Each chapter is followed by a list of the pertinent references, where the additional material about the work referred to may be found.

I am grateful to many people who directly or indirectly influenced me to undertake this project. Professor Benjamin M. Gimarc (Columbia, S.C.) convinced me that I should write this book in the light of the role of the Zagreb Group in the remarkable growth of the chemical applications of graph theory in the last decade. He was kind enough to read the entire book in the manuscript form and to give useful comments leading to the improvement of the book in every way. This is a good opportunity to thank him for critical, but friendly discussions, comments, and help.

P. Křivka (Pardubice), I. Mladenov (Sofia), and B. Mohar (Ljubljana) also commented on several chapters of the book. I am thankful to them for their kind help.

I wish to thank several people for helping me to enter the fascinating field of (chemical) graph theory. I am indebted to Professor A. T. Balaban (Bucharest), Professor F. Harary (Ann Arbor), and Professor A. J. Schwenk (Annapolis) for teaching me the elements of graph theory. I am also thankful to Professor W. C. Herndon (El Paso), Professor B. A. Hess, Jr. (Nashville), Professor H. Hosoya (Tokyo), Professor L. Klasinc (Zagreb), Dr. R. B. Mallion (Canterbury), Professor O. E. Polansky (Mülheim/Ruhr), Professor V. Prelog (Zürich), Professor M. Randić (Ames), and Professor L. J. Schaad (Nashville) for encouragement and support over the years.

Finally, I wish to thank Dr. B. Džonova-Jerman-Blažić (Ljubljana), Dr. I. Gutman (Kragujevac), Dr. M. Milun (Zagreb), and Dr. P. Ilic (Sarajevo), my former graduate students, for their enthusiasm and hard work, and the past and present members, and the guests, of the Zagreb Group: Dr. S. Bosamac, Dr. D. Bonchev (Burgas), Mrs. M. Barysz (Katowice), Dr. A. Graovac, Mr. Ž. Jeričević, Mr. A. Jurić (Banja Luka), Mr. D. Kasum, Professor J. V. Knop (Düsseldorf), Dr. A. Sabljić, Dr. J. Seibert (Hradec Králové), Dr. A. Velenik-Oliva, Professor C. F. Wilcox, Jr. (Ithaca), and Dr. T. Živković for their patient collaboration during the past 14 years.

REFERENCES

1. **Higgins, W.,** *A Comparative View of the Phlogistic and Anti-Phlogistic Theories,* Murray, London, 1789.
2. **Kopp, H.,** *Ann. Chem.,* 41, 79, 1842; 41, 169, 1842.
3. **Crum Brown, A.,** *Trans. R. Soc. Edinburgh,* 23, 707, 1864.
4. **Laurent, A.,** *Ann. Chim. Phys.,* 18, 266, 1864.
5. **Cayley, A.,** *Phil. Mag.,* 18, 374, 1859.
6. **Cayley, A.,** *Phil. Mag.,* 47, 444, 1874.
7. **Cayley, A.,** *Chem. Ber.,* 8, 1056, 1875.
8. **Flavitzky, F.,** *J. Russ. Chem. Soc.,* 160, 1871.
9. **Sylvester, J. J.,** *Nature (London),* 17, 284, 1877—1878; *Am. J. Math.,* 1, 64, 1878.
10. **Biggs, N. L., Lloyd, E. K., and Wilson, R. J.,** *Graph Theory 1736—1936,* Clarendon, Oxford, 1976.
11. **Rouvray, D. H.,** *R. I. C. Rev.,* 4, 173, 1971.
12. **Mallion, R. B.,** *Chem. Br.,* 9, 242, 1973.
13. **Gutman, I. and Trinajstić, N.,** *Topics Curr. Chem.,* 42, 49, 1973.
14. **Wilson, R. J.,** in *Colloquia Mathematica Societatis János Bolyai,* Vol. 18 Combinatorics, Kezthely (Hungary), 1976, 1147.
15. **Cvetković, D., Doob, M., and Sachs, H.,** *Spectra of Graphs,* Academic Press, New York, 1980.

THE AUTHOR

Nenad Trinajstić is a Senior Researcher at the Rugjer Bošković Institute in Zagreb, Croatia, Yugoslavia and a Professor at the Department of Chemistry, Faculty of Science and Mathematics, University of Zagreb. He received the B.Sc. (1960), M.Sc. (1966), and Ph.D. (1967) degrees from the University of Zagreb. During the period 1964 to 1966, he was a predoctoral fellow with Prof. John N. Murrell at the University of Sheffield and the University of Sussex, U.K. and collaborated with him in research on MO interpretation of electronic spectra of conjugated molecules and on the development of criteria for producing localized orbitals. Dr. Trinajstić's postdoctoral years (1968 to 1970) were spent with Prof. Michael J. S. Dewar at the University of Texas at Austin. The main research interest during this period was the development of a convenient semiempirical MO theory for studying the ground states of large molecules. After returning to Zagreb from the U.S., he initiated research on chemical applications of graph theory and has greatly contributed to the revival of the uses of graph theory in chemistry. In the last decade, he spent some time at various universities such as the University of Trieste, Italy (collaborating with Prof. Vinicio Galasso), the University of Utah, Salt Lake City (collaborating with Prof. Frank E. Harris), the University of Oxford, U.K. (collaborating with Dr. Roger B. Mallion), the University of South Carolina, Columbia (collaborating with Prof. Benjamin M. Gimarc), the University of Düsseldorf, West Germany (collaborating with Prof. Jan V. Knop), the Iowa State University, Ames (collaborating with Prof. Milan Randić), Higher School of Chemical Technology, Burgas, Bulgaria (collaborating with Dr. Danail Bonchev), and the University of Sussex, Brighton, England (collaborating with Prof. John N. Murrell). He had a number of B.Sc., M.Sc., and Ph.D. students, and has published over 200 papers in the fields of Theoretical Organic Chemistry and Mathematical Chemistry. His present main research interest is the chemical applications of graph theory. In 1972, he received the City of Zagreb Award for research. In 1982, he received The Rugjer Bošković Award for his work in chemical graph theory. He is married (Judita), has two children (daughter: Regina and son: Dean), and a grandson (Sebastijan).

TABLE OF CONTENTS

Volume I

Chapter 8
Subspectral Molecules

TABLE OF CONTENTS

Volume II

Chapter 1

TOPOLOGICAL RESONANCE ENERGY

The resonance energy, RE, is a quantity which is being used for predicting aromaticity in conjugated structures.[1] By aromaticity we mean the intuitive, but practical, concept for predicting and characterizing the stability of conjugated species.[2-5] The general definition of resonance energy is as follows,

$$RE = E_\pi \text{ (conjugated molecule)} - E_\pi \text{ (reference structure)} \qquad (1)$$

where E_π (conjugated molecule) and E_π (reference structure) are the π-electron energies of a given conjugated molecule and the corresponding hypothetical reference structure, respectively, usually obtained by means of Hückel theory. Since the reference structure is a nonexisting entity, its choice is to a great extent arbitrary. There are many proposals, given in the literature,[6] of ways to select the reference structure. The first proposal, the classical Hückel resonance energy, HRE,[7] is based on the reference structure containing $n_{C=C}$ carbon-carbon double bonds isolated by conjugation barriers,

$$HRE = E_\pi \text{ (conjugated molecule)} - 2n_{C=C} \qquad (2)$$

Note, that we use the normalized form of Hückel theory, e.g., the parameter α for the carbon atom is taken as the zero-point energy, $\alpha_C = 0$, and the bond parameter β as the energy unit, $\beta_{CC} = 1$. The HRE criterion is shown to fail in many cases because rather unstable molecules are predicted to be aromatic on the grounds of their HREs being large (see Table 1).

For example, we note that the HRE (per π-electron) values of benzene (0.33) and pentalene (0.31) are comparable, thus, indicating the latter compound to be as aromatic as benzene in spite of the fact that pentalene is a very unstable molecule.[8] Similarly, heptalene, fulvene, heptafulvene, and fulvalene are all predicted to be aromatic on the grounds of their large HRE values.

Such predictions have been largely disproved by efforts to synthesize these compounds.[8,9] Therefore, the HRE quantity cannot be used as a reliable index for classifying aromatic structures. The origin of the problems with HRE can be traced to the fact that HRE is grossly proportional to N (the number of π-electrons in the systems).[10] Since,

$$n_{C=C} = N/2 \qquad (3)$$

it follows,

$$HRE = E_\pi \text{ (conjugated molecule)} - N \qquad (4)$$

Now, if we use the approximate McClelland-type formula[11] for π-energy of a conjugated system, the following relation is obtained,

$$E_\pi \text{ (conjugated molecule)} \approx a \left[2N(N + R - 1)\right]^{1/2} \qquad (5)$$

Using relation (5) and utilizing the reasonable assumption N>>R−1 one obtains,

$$HRE \approx N(2^{1/2} a - 1) \qquad (6)$$

or for the McClelland value of a (0.92),[11]

$$HRE \approx 0.3 N \qquad (7)$$

Table 1
HÜCKEL RESONANCE ENERGIES,
HREs, OF SOME CONJUGATED
MOLECULES

Compound	HRE[a]	HRE(PE)[b]
Benzene	2.000	0.333
Naphthalene	3.683	0.368
Anthracene	5.314	0.380
Phenanthrene	5.448	0.389
Perylene	8.245	0.412
Coronene	10.572	0.481
Pyrene	6.505	0.407
Pentalene	2.456	0.307
Heptalene	3.618	0.302
Fulvene	1.466	0.244
Heptafulvene	1.994	0.249
Fulvalene	2.799	0.280

[a] HRE values are calculated using the Hückel values taken from Coulson, C. A. and Streitwieser, A., Jr., *Dictionary of π-Electron Calculations*, Pergamon Press, Oxford, 1965.
[b] HRE(PE) = HRE/N, where N is the number of π-electrons in the molecule.

Several attempts have been made to redefine RE in order to obtain the better agreement between the theory and experiment.[12] An important improvement was introduced by Dewar[13] who proposed the use of an acyclic polyene-like reference structure in RE calculations. This novel type of resonance energy is named Dewar resonance energy, DRE.[14] DRE is defined as,

$$DRE = E_\pi \text{ (conjugated molecule)} - \sum_{j=1}^{L} n_j E_j \qquad (8)$$

where n_j and E_j represent the number of particular polyene bond types and the corresponding energy parameters, respectively. Dewar and co-workers[15-19] obtained good predictions of aromaticity for many conjugated molecules using this method (see Table 2).

However, Dewar, in his work, has used an original version of the SCF π-MO theory, parametrized to reproduce the ground state properties of conjugated molecules. This lead some people to believe that the change of theory from non-SCF to SCF level is responsible for the improvement of the predictions.

Subsequent studies have shown that this is not the case. The fundamental step leading to good predictions of aromatic stability was not the use of more sophisticated MO theory, but rather use of the acyclic polyene-like reference structure.[20] That was confirmed completely by Hess and Schaad,[21] and the Zagreb Group[22] who simply transplanted the Dewar definition of the reference structure from SCF π-MO theory to Hückel theory.

The Zagreb Group[22] reported the calculation of Dewar resonance energies in which the acyclic polyene-like reference structure was presented by a two-bond parameters scheme ($E_{C-C} = 0.5$, $E_{C=C} = 2.00$), while Hess and Schaad[21] used an eight-bond parameters scheme ($E_{C-C} = 0.4358$, $E_{HC-C} = 0.4362$, $E_{HC-CH} = 0.4660$, $E_{C=C} = 2.1716$, $E_{HC=C} = 2.1083$, $E_{H_2C=C} = 2.0000$, $E_{HC=CH} = 2.0699$, $E_{H_2C=CH} = 2.0000$) to approximate the localized reference structure. It is worth mentioning that Schaad and Hess[23] investigated the origin of the reference structures with the two-bond and the eight-bond parameters. They have treated acyclic reference structures as collections of perturbed ethylene molecules. The perturbation

Table 2
DEWAR RESONANCE ENERGIES (IN eV)
OF SOME CONJUGATED MOLECULES

| Molecule | Calculated Dewar resonance energy | | Status[c] |
	DRE[a]	DRE(PE)[b]	
Cyclobutadiene	−0.78	−0.20	AA
Benzene	0.87	0.15	A
Naphthalene	1.32	0.13	A
Anthracene	1.60	0.11	A
Phenanthrene	1.93	0.14	A
Perylene	2.62	0.13	A
Pyrene	1.82	0.11	A
Pentalene	0.006	0.001	NA
Heptalene	0.09	0.008	NA
Fulvene	0.05	0.008	NA
Cyclooctatetraene	−0.11	−0.01	NA
[18]-Annulene	0.126	0.007	NA
Acenaphthylene	1.34	0.12	A

[a]   Data from Dewar, M. J. S. and de Llano, C., *J. Am. Chem. Soc.*, 91, 789, 1969; Trinajstić, N., *Rec. Chem. Prog.*, 32, 85, 1971.

[b]   DRE(PE) = DRE/N; where N is the number of π-electrons in the conjugated molecule.

[c]   A = aromatic, NA = nonaromatic, AA = antiaromatic.

Table 3
COMPARISON OF BOND ENERGIES OBTAINED
BY THE PERTURBATION TREATMENT THROUGH
THE FOURTH ORDER WITH EMPIRICAL BOND
ENERGIES

Bond	Empirical bond energy	Perturbation bond energy
C–C	0.4358	0.4063
HC–C	0.4362	0.4375
HC–CH	0.4660	0.4688
C=C	2.1716	2.2500
HC=C	2.1083	2.1250
H_2C=C	2.0000	2.0000
HC=CH	2.0699	2.0625
H_2C=CH	2.0000	2.0000

treatment through the second order produced the two-bond parameter scheme, while the perturbation treatment through the fourth order produces the eight-bond parameter scheme with the bond energies only slightly differing from the empirical ones earlier determined by Hess and Schaad[21] (see Table 3).

For many compounds the two DRE-like methods give very similar results, but there are exceptions due to the fact that a two-bond energy parameter scheme rather poorly approximates π-energies of branched polyenes. However, both methods correctly differentiate aromaticity in benzenoid and nonbenzenoid hydrocarbons (see Table 4).

Detailed analysis of DRE and DRE-like methods revealed several limitations of these calculations: results depend on the parametric values chosen for the reference bond energies,

Table 4
RESONANCE ENERGIES OF SOME CONJUGATED
HYDROCARBONS CALCULATED USING THE ZAGREB
GROUP MODEL AND THE HESS-SCHAAD MODEL

	Calculated resonance energy				
Molecule	A_s[a]	$A_s(PE)$[b]	RE[c]	RE(PE)[d]	Status[e]
Benzene	0.440	0.073	0.39	0.065	A
Naphthalene	0.563	0.056	0.55	0.055	A
Anthracene	0.634	0.045	0.66	0.047	A
Phenanthrene	0.768	0.055	0.77	0.055	A
Chrysene	0.952	0.053	0.96	0.053	A
Perylene	0.965	0.048	0.97	0.049	A
Pyrene	0.785	0.049	0.81	0.051	A
Azulene	0.244	0.024	0.23	0.023	A
Pentalene	−0.144	−0.018	−0.14	−0.018	AA
Heptafulvene	−0.086	−0.011	−0.02	−0.003	NA
Fulvalene	−0.321	−0.032	−0.60	−0.060	AA
Heptalene	−0.022	−0.002	−0.05	−0.004	NA
Fulvene	−0.094	−0.016	−0.01	−0.002	NA
Heptafulvalene	−0.155	−0.011	−0.20	−0.014	AA
Cyclobutadiene	−1.040	−0.260	−1.07	−0.268	AA

[a]   Data from Milun, M., Sobotka, Ž., and Trinajstić, N., *J. Org. Chem.*, 37, 139, 1972.
[b] $A_s(PE) = A_s/N$, where N is the number of π-electrons in the conjugated system.
[c]   Data from Hess, B. A., Jr. and Schaad, L. J., *J. Am. Chem. Soc.*, 93, 305, 1971.
[d] $RE(PE) = RE/N$, where N is the number of π-electrons in a molecule.
[e]   A = aromatic, NA = nonaromatic, AA = antiaromatic.

the number of the bond parameters increases rapidly when heteroatoms are considered, DRE values can be obtained only for molecules possessing classical structures, thus excluding the number of nonclassical conjugated systems, i.e., conjugated ions and radicals, and homo-aromatic structures (However, recently Hess and Schaad[24] have extended their model to embrace conjugated ions and radicals by introducing a novel reference structure. This extension requires 11 additional parameters). One would like to have DRE without these shortcomings. A search for the DRE-like quantity, free of the above limitations, produced an index independent of bond parameters and is directly related to the topology of the molecular π-network. It is named *topological resonance energy*, TRE, and is obtained by translating directly the DRE concept into the formalism of chemical graph theory.[25] Thus, TRE is not really a new aromaticity index, but an optimal nonempirical DRE-like quantity, which appears to be applicable to a variety of conjugated systems.

I. DERIVATION OF TOPOLOGICAL RESONANCE ENERGY

In this section, the theory behind the TRE concept will be outlined. Attention will be placed on the reference structure. The aim is to obtain an acyclic polyene-like reference structure for which construction of all substructural details of a conjugated molecule, except the cycles, should be taken into account. The target is such a reference structure because we relate the aromatic stability in a conjugated molecule with the contributions from cycles to the total π-electron energy of a system. This was achieved by the convenient adaptation[25] of the Sachs formula[26] for the coefficients of the characteristic polynomial: the cycles are left out of the formula (23) in Chapter 5, Volume I.

$$a_n^{ac}(G) = \sum_{s \in S_n^{ac}} (-1)^{c(s)} \qquad (9)$$

This relation considers only the complete set of acyclic Sachs graphs, S_n^{ac}, of a graph (molecule), or in another words for construction of a_n^{ac} coefficients one needs to take into account only K_2 components. We call a Sachs graph[27] s acyclic if $r(s) = 0$, and thus,

$$S_n^{ac} = \left\{ s \in S_n \mid r(s) = 0 \right\} \tag{10}$$

Therefore, the polynomial constructed by means of relation (9) contains coefficients free of any cyclic contributions and represents the optimal approximation to the acyclic reference structure,

$$P^{ac}(G;x) = \sum_{n=0}^{N} a_n^{ac}(G) x^{N-n} \tag{11}$$

Polynomial (11) is a crucial concept in the topological theory of aromaticity and is named the *acyclic polynomial*.[25] (Some authors[28] disagree with this statement by saying that the main objective is to obtain the roots of the polynomials, which correspond to energy levels of a molecule and a reference structure. However, the problem is not in obtaining the roots of a polynomial, but in setting up a structure which will lead to polynomials and roots, and to resonance energies which in turn will correlate well with the experimental measures of aromaticity). Sometimes the acyclic polynomial is also referred to as the *reference polynomial*.[29] In the mathematical literature, the acyclic polynomial is called the *matching polynomial*.[30]

Generally, it is not possible to associate a graph with the acyclic polynomial. However, it is possible for some simple π-electronic structures to give the *acyclic graphs*. Acyclic graph G^{ac}, corresponding to a graph G, is defined as the graph whose characteristic polynomial $P(G^{ac}; x)$ equals the acyclic polynomial $P^{ac}(G; x)$ of G.

Example

Hückel graphs (G) Corresponding acyclic graph (G^{ac})

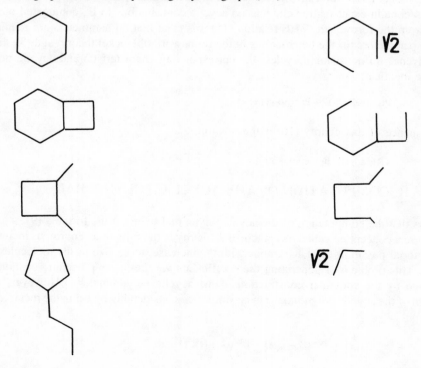

The acyclic graphs given above are the edge-weighted graphs, G_{EW}, with the heterobond parameter k equal to $\sqrt{2}$ or $\sqrt{3}$.

The solutions of the acyclic polynomial x_j^{ac} ($j = 1, 2 \ldots, N$) represent the complete set of (Hückel) energies of the acyclic reference structure.

$$E_\pi \text{ (acyclic reference structure)} = \sum_{j=1}^{N} h_j x_j^{ac} \qquad (12)$$

where h_j is the occupancy number of energy levels associated with the reference structure.

Introducing the expression for the total π-electron energy of a conjugated molecule,

$$E_\pi \text{ (conjugated molecule)} = \sum_{j=1}^{N} g_j x_j \qquad (13)$$

and Equation (12) in (1) the relation for the topological resonance energy is obtained in a rather simple form,

$$TRE = \sum_{j=1}^{N} (g_j x_j - h_j x_j^{ac}) \qquad (14)$$

which for case when $g_j = h_j$ reduces to elegant and condensed expression,

$$TRE = \sum_{j=1}^{N} g_j (x_j - x_j^{ac}) \qquad (15)$$

The application of relation (15) is straightforward. What one needs are Hückel energies x_j and the roots of the acyclic polynomial x_j^{ac}.

In order to ensure that TRE values are always real, it is necessary to prove that all roots of an acyclic polynomial are real. The theorem which assures that all roots of $P^{ac}(G; x)$ are real is available[31] and will be discussed later on in this chapter.

The TRE model has several favorable features in comparison with DRE-like methods: (1) TRE values contain exactly all cyclic and no acyclic contributions to E_π(conjugated molecule); (2) from the form of the TRE relation (15) it is clear that no parameters are required outside the parameters for the heteroatoms in the framework of Hückel theory; and (3) since acyclic polyenes do not contain cycles, the corresponding characteristic and acyclic polynomials are identical,

$$P(\text{polyene}; x) \equiv P^{ac}(\text{polyene}; x) \qquad (16)$$

The consequence of the identity (16) is the relation,

$$TRE(\text{acyclic polyene}) = 0 \qquad (17)$$

II. COMPUTATION OF THE ACYCLIC POLYNOMIAL

It appears that the computation of the acylic polynomial by using the formula (9), which has shown great conceptual value, has practical difficulty in the form of an enormous increase of combinatorial possibilities of K_2 graphs with the increase of the size of the molecules to be studied. This problem of generating the coefficients via the Sachs procedure cannot be avoided even by the computer construction of the acyclic polynomial.[32,33] However, the computation of the acyclic polynomial can be simplified considerably by using the recurrence relation,[34,35]

$$P^{ac}(G; x) = P^{ac}(G - e; x) - P^{ac}(G - (e); x) \qquad (18)$$

where G-e denotes the subgraph of a graph G obtained by removing the edge *e* from G and G-(e) is the subgraph obtained by deleting the edge *e* and its incident vertices,

Example

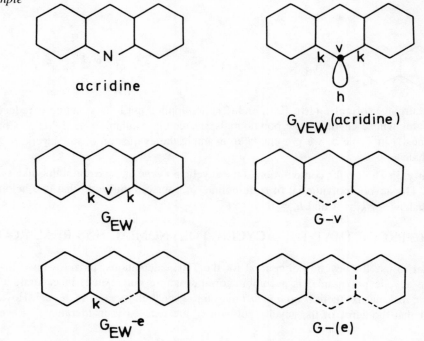

G   G-e   G-(e)

Equation (18) is a generalization of the Heilbronner formula,[36]

$$P(G; x) = P(G - e; x) - P(G - (e); x) \qquad (19)$$

which is valid for acyclic polyenes only.

The computation of the acyclic polynomial for heterocyclic structures is also based on the Equation (18) which is modified[34,35] for this case to include the Hückel parameters for heteroatoms *h* and *k*,

$$P^{ac}(G_{VEW}; x) = P^{ac}(G_{EW}; x) - h\, P^{ac}(G - v; x) \qquad (20)$$

$$P^{ac}(G_{EW}; x) = P^{ac}(G - e; x) - k^2\, P^{ac}(G - (e); x) \qquad (21)$$

where G_{VEW} is a vertex- and edge-weighted graph with a loop of a weight *h* attached to the vertex *v*, whereas G_{EW} is an edge-weighted graph with an edge *e* of a weight *k*. G_{EW} and G-v are subgraphs obtained by deleting the loop at the vertex *v* from G_{VEW} and by deleting the vertex *v* and incident edges from G_{EW}, respectively.

Example

Table 5

THE CONSTRUCTION OF THE ACYCLIC POLYNOMIAL OF NAPHTHALENE

G(naphthalene)

$$P^{ac}\left(\text{[naphthalene with edge e]} ; x \right) = P^{ac}\left(\text{[naphthalene]} ; x \right) - P^{ac}\left(\text{[naphthalene with dashed bonds]} ; x \right)$$

$$P^{ac}\left(\text{[structure]} ; x \right) \equiv L_4 L_4$$

$$P^{ac}\left(\text{[structure with e]} ; x \right) = P^{ac}\left(\text{[structure]} ; x \right) - P^{ac}\left(\text{[structure]} ; x \right)$$

$$P^{ac}\left(\text{[chain structure]} ; x \right) \equiv L_{10}$$

$$P^{ac}\left(\text{[chain structure]} ; x \right) \equiv L_8$$

$$P^{ac}\left(\text{[naphthalene]} ; x \right) = L_{10} - L_8 - L_4 L_4 = x^{10} - 11 x^8 + 41 x^6 - 61 x^4 + 31 x^2 - 3$$

The recurrence formulae (18), (20), and (21) are applied until the structure is reduced to linear chains whose characteristic polynomials are readily available (see Table 3, Chapter 5, Volume I). In Table 5, we present as an example the evaluation of the acyclic polyene of naphthalene.

Similarly, in Table 6 the computation of the acyclic polyene of the α-quinoline-like system is given. The acyclic polynomial of α-quinoline, of course, reduces to that one belonging to naphthalene for $k = 1$ and $h = 0$.

III. THE PROOF THAT THE ACYCLIC POLYNOMIAL HAS REAL ROOTS

In order to use the acyclic polynomial for the TRE calculations, its zeros must be real. Since the acyclic polynomial is a combinatorial structure not related, in general, to any symmetric real matrix, there is no *a priori* reason to expect all its roots to be real. However, the proof that the zeros of the acyclic polynomial are real exists in literature.[31] Recently,

Table 6
THE CONSTRUCTION OF THE ACYCLIC POLYNOMIAL OF α-QUINOLINE-LIKE SYSTEM

$$\alpha\text{-quinoline}$$

$$G_{VEW}$$

$$P^{ac}\left(\quad;x\right) = P^{ac}\left(\quad;x\right) - h\,P^{ac}\left(\quad;x\right)$$

$$P^{ac}\left(\quad;x\right) = P^{ac}\left(\quad;x\right) - k^2 P^{ac}\left(\quad;x\right)$$

$$P^{ac}\left(\quad;x\right) = P^{ac}\left(\quad;x\right) - k^2 P^{ac}\left(\quad;x\right)$$

$$P^{ac}\left(\quad;x\right) = P^{ac}\left(\quad;x\right) - P^{ac}\left(\quad;x\right)$$

$$P^{ac}\left(\quad;x\right) \equiv L_9 L_1$$

$$P^{ac}\left(\quad;x\right) \equiv L_4 L_3 L_1$$

$$P^{ac}\left(\quad;x\right) \equiv L_8$$

$$P^{ac}\left(\quad;x\right) = P^{ac}\left(\quad;x\right) - P^{ac}\left(\quad;x\right)$$

Table 6 (continued)

THE CONSTRUCTION OF THE ACYCLIC POLYNOMIAL OF α-QUINOLINE-LIKE SYSTEM

$$P^{ac}\left(\begin{array}{c}\end{array}; x\right) \equiv L_8$$

$$P^{ac}\left(\begin{array}{c}\end{array}; x\right) \equiv L_4 L_2$$

$$P^{ac}\left(\begin{array}{c} e\end{array}; x\right) = P^{ac}\left(\begin{array}{c}\end{array}; x\right) - P^{ac}\left(\begin{array}{c}\end{array}; x\right)$$

$$P^{ac}\left(\begin{array}{c}\end{array}; x\right) \equiv L_9$$

$$P^{ac}\left(\begin{array}{c}\end{array}; x\right) \equiv L_4 L_3$$

$$P^{ac}\left(\begin{array}{c} v \\ k \quad k \\ h\end{array}; x\right) = L_9 L_1 - L_4 L_3 L_1 - k^2\left\{2L_8 - L_2 L_4\right\} - h\left\{L_9 - L_3 L_4\right\}$$

$$= x^{10} - 9x^8 + 26x^6 - 27x^4 + 7x^2 - k^2\left\{2x^8 - 15x^6 + 34x^4 - 24x^2 + 3\right\} - h\left\{x^9 - 9x^7 + 26x^5 - 27x^3 + 7x\right\}$$

the reality of the zeros of the acyclic polynomial, belonging to chemical structures has been independently discussed by several groups.[37-41] Among these is the very interesting work by Gutman[39] who has elegantly demonstrated that the acyclic polynomial of heteroconjugated molecules (represented by vertex- and edge-weighted graphs) has real zeros. Here we will review the Heilmann-Lieb theorem.[31]

Heilmann and Lieb stated that if G is a complete graph, then the zeros of $P^{ac}(G; x)$ are all real and, moreover, if v is any vertex of G then the zeros of $P^{ac}(G; x)$: x_1, x_2, . . . , x_N, and the zeros of P^{ac} (G-v; x): x_1', x_2', . . . , x_{N-1}' obey a strict interlacing relationship,

$$x_1 < x_i' < x_{i+1} \tag{22}$$

for all $i = 1, 2, . . . , N-1$. If G contains only one vertex the above result is obviously true. The statement (22) can be proved for $N > 1$ by induction, assuming that (22) holds for all complete graphs G' with $N(G') \leqslant N-1$.

There are several recurrence relations available for the construction of acyclic polynomials.[34,42,43] One relation has already been discussed in Section II. Here we will use another one which is more convenient for the present discussion,

$$P^{ac}(G; x) = x\, P^{ac}(G - v_r; x) - \sum_s P^{ac}(G - v_r - v_s; x) \qquad (23)$$

where G-v_r is a subgraph (with $N - 1$ vertices) obtained by deleting vertex r and incident edges from G, while G-v_r-v_s is a subgraph (with $N - 2$ vertices) obtained by removing adjacent vertices r and s and the incident edges from G.

Example

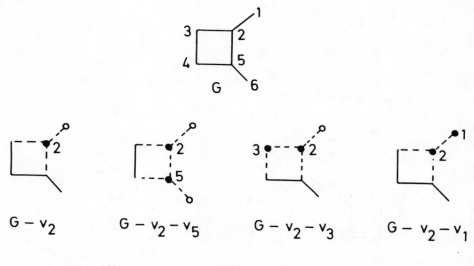

$$P^{ac}(G; x) = x\, P^{ac}(G - v_2; x) - \left\{ P^{ac}(G - v_2 - v_5; x) + P^{ac}(G - v_2 - v_3; x) + \right.$$

$$\left. P^{ac}(G - v_2 - v_1; x) \right\} = x\, [x(x^4 - 3x^2 + 1) - x^2(x^2 - 1) +$$

$$x(x^3 - 2x) + x^4 - 3x^2 + 1] = x^6 - 6x^4 + 7x^2 + 1$$

$P^{ac}(G$-$v_r; x)$ is a polynomial of degree $N - 1$, while the sum polynomial $\sum_s P^{ac}(G$-v_r-$v_s; x)$ of degree $N - 2$. From the induction assumption it follows that the zeros of $P^{ac}(G$-v_r-$v_s; x)$ interlace the zeros of $P^{ac}(G$-$v_r; x)$ for all vertices s of the subgraph G-v_r. Now, an analysis which makes use of characteristic behavior of polynomials with regard to their zeros,[44] immediately reveals that between each pair zeros belonging to the polynomial $P^{ac}(G$-$v_r; x)$ there will be *exactly* one zero of $P^{ac}(G; x)$. In addition, $P^{ac}(G; x)$ will have *exactly* one zero on the left from the first zero of $P^{ac}(G$-$v_r; x)$ and *exactly* one zero on the right from the last zero of $P^{ac}(G$-$v_r; x)$. From this follows that $P^{ac}(G; x)$ must have *exactly* N real zeros, because $P^{ac}(G; x)$ is a polynomial of degree N is unknown x and thus it cannot have any other zeros. This result will be illustrated below by means of a numerical analysis.

Example

$$P^{ac}(G; x) = x^6 - 6x^4 + 7x^2 - 1$$

$$\text{zeros: } \{2.119, 1.159, 0.407, -0.407, -1.159, -2.119\}$$

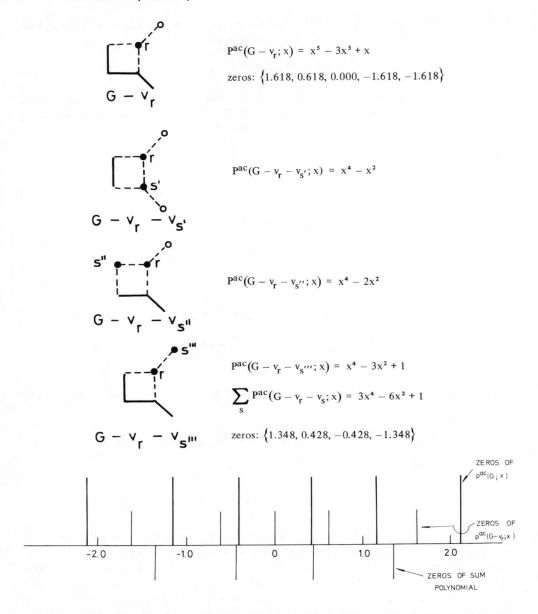

$$P^{ac}(G - v_r; x) = x^5 - 3x^3 + x$$

zeros: $\{1.618, 0.618, 0.000, -1.618, -1.618\}$

$$P^{ac}(G - v_r - v_{s'}; x) = x^4 - x^2$$

$$P^{ac}(G - v_r - v_{s''}; x) = x^4 - 2x^2$$

$$P^{ac}(G - v_r - v_{s'''}; x) = x^4 - 3x^2 + 1$$

$$\sum_s P^{ac}(G - v_r - v_s; x) = 3x^4 - 6x^2 + 1$$

zeros: $\{1.348, 0.428, -0.428, -1.348\}$

The statement (22) may be extended to embrace graphs which are not complete. Heilmann and Lieb[31] summarize this in the following theorem: For any graph G, the zeros of $P^{ac}(G; x)$ are all real. Furthermore, if v is any vertex of G and if $x_1, x_2, \ldots, x_N$ are the zeros of $P^{ac}(G; x)$, while the zeros of $P^{ac}(G-v; x)$ are $x_1', x_2', \ldots, x_{N-1}'$, then the interlacing relationship,

$$x_i \leq x_i' \leq x_{i+1} \tag{24}$$

holds for all $i = 1, 2, \ldots, N-1$.

A question could be asked when the inequalities (24) will be strict, because the strict interlacing means that the zeros are mutually distinct. It is established[31] that all the inequalities in (24) are strict if G contains a Hamiltonian path that ends at the vertex v.

IV. CONNECTION BETWEEN THE CHARACTERISTIC POLYNOMIAL AND THE ACYCLIC POLYNOMIAL

The characteristic polynomial and the acyclic polynomial of G are related in a simple way. For acyclic polyenes these two polynomials are identical; see Equation (16). In the case of annulenes the difference between $P(C_N; x)$ and $P^{ac}(C_N; x)$ is the rather plain,

$$P(C_N; x) - P^{ac}(C_N; x) = 2 \tag{25}$$

where C_N is a short hand notation of N-cycle. If we write explicitly $P(C_N; x)$ and $P^{ac}(C_N; x)$, the above expression immediately arises,

$$P(C_N; x) = \sum_{k=0}^{[N/2]} (-1)^k \, p(C_N; k) \, x^{N-2k} - 2 \tag{26}$$

$$P^{ac}(C_N; x) = \sum_{k=0}^{[N/2]} (-1)^k \, p(C_N; k) \, x^{N-2k} \tag{27}$$

where $p(C_N; k)$ is the number of k-matching in one N-cycle. A k-matching in a graph G is a selection of $2k$ vertices which are pairwise joined by k edges.[45]

The connection between $P(G; x)$ and $P^{ac}(G; x)$ for polycyclic systems is more complicated,[46]

$$P(G; x) - P^{ac}(G; x) = -2 \sum_m P^{ac}(G - C_m; x) + 4 \sum_{m<n} P^{ac}(G - C_m - C_n; x) -$$

$$8 \sum_{m<n<p} P^{ac}(G - C_m - C_n - C_p; x) + \ldots \tag{28}$$

where the summations go over all pairs, triplets, . . . , etc. of mutually disconnected cycles which are contained in G. $G-C_m$, $G-C_m-C_n$, $G-C_m-C_n-C_p$, . . . , etc. are subgraphs obtained by removing successively C_m; C_m and C_n; C_m, C_n, and C_p, . . . ; and the adjacent bonds from G.

Example

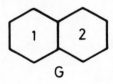

$$G - C_6^1 = L_4 \qquad G - C_6^2 = L_4 \qquad G - C_{10} = L_o$$

$$P\left(\includegraphics{}; x\right) - P^{ac}\left(\includegraphics{}; x\right) = -2(L_4 + L_4 + L_o)$$

$$= -4x^4 + 12x^2 + 6$$

The relation (28) may be used for studying the effect of individual cycles on TRE and E_π.[47] It has been shown that in polycyclic conjugated systems the $(4m)$-membered cycles always *destabilize* conjugated molecule. On the other hand, the $(4m + 2)$-cycles *stabilize* conjugated structures, though examples have been found in which $(4m + 2)$-rings have a destabilizing effect on the molecular stability.[48] That contribution to E_π (conjugated molecule) which comes from the presence of the cycle C_i (i = 3, 4, 5, . . .) is called the *cycle energy*,[48] E_{cycle}, and is given by the following energy difference,

$$E_{cycle} = E_\pi(\text{conjugated molecule}) - E(G - C_i) \qquad (29)$$

where $E(G-C_i)$ stands for the π-electron energy of the stucture obtained when the cycle C_i is removed from the molecule.[49]

Example

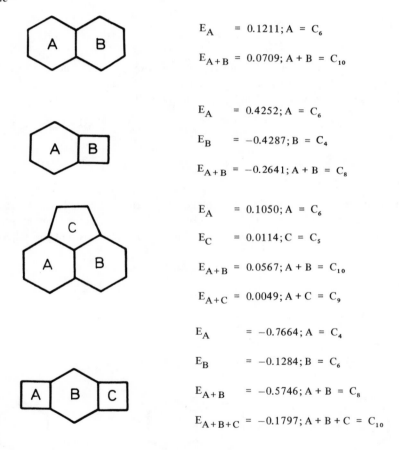

$E_A \quad = 0.1211; A = C_6$

$E_{A+B} = 0.0709; A + B = C_{10}$

$E_A \quad = 0.4252; A = C_6$

$E_B \quad = -0.4287; B = C_4$

$E_{A+B} = -0.2641; A + B = C_8$

$E_A \quad = 0.1050; A = C_6$

$E_C \quad = 0.0114; C = C_5$

$E_{A+B} = 0.0567; A + B = C_{10}$

$E_{A+C} = 0.0049; A + C = C_9$

$E_A \quad = -0.7664; A = C_4$

$E_B \quad = -0.1284; B = C_6$

$E_{A+B} \quad = -0.5746; A + B = C_8$

$E_{A+B+C} = -0.1797; A + B + C = C_{10}$

These results serve as a quantitative test of the Hückel rule. The Hückel rule was originally proposed for annulenes only.[50] Later it was conjectured that it holds for any arbitrary conjugated molecule.[51-53] However, these authors, and many others,[1,4,7] have been using the Hückel rule as a *qualitative* rule, though it was quantified for the monocycles.[54] Thus, the Hückel rule was used only to indicate whether a certain cycle has stabilizing or destabilizing effect on the molecular stability without the knowledge of the magnitude of this effect. The above quantitative approach has produced an important result not obtainable by the qualitative studies: in some cases the Hückel $(4m + 2)$ rule is violated.

V. NORMALIZATION OF TRE

TRE values can be calculated using Equations (14) and (15) and can be used for comparing the aromatic stabilities of isomeric structures. However, for molecules of various sizes the TRE values only cannot be used meaningfully. For example, the TRE values of benzene (0.276) and naphthalene (0.390) cannot be compared because these two molecules have a different size (i.e., a different number of atoms and bonds). In order to avoid the *size effect*,[55] the TRE must be normalized.

TRE may be normalized in several ways to produce the intrinsic topological resonance quantity.[56] Here we report two normalization procedures. One is to take into account the total number of π-electrons in the molecule, N, which leads to the TRE (per π-electron), TRE(PE), value,[34]

$$TRE(PE) = TRE/N \qquad (30)$$

The second procedure of normalization with respect to the ring-like components,[56,57] since TRE contains only cyclic contributions to E_π, leads to the quantity TRE(per ring bond), TRE(PRB),

$$TRE(PRB) = TRE/RB \qquad (31)$$

where RB is the number of bonds within a given ring. The TRE(PRB) normalization ignores all side chains as well as the actual number of π electrons in the system. However, the comparison between TRE(PE) and TRE(PRB) reveals that both normalizations lead to the predictions of the same quality.[56,58] Therefore, TRE(PE) will be ordinarily used as a criterion for classifying conjugated structures. Molecules with large positive value of TRE(PE) are considered *aromatic*, those with TRE(PE) values close to zero *nonaromatic*, and finally compounds with large negative values of TRE(PE) *antiaromatic*. The practical threshold values recommended to be used in the interpretation of the results and the classification are the following:

1. Aromatic molecules: $TRE(PE) > 0.01$
2. Nonaromatic molecules: $-0.01 < TRE(PE) < 0.01$
3. Antiaromatic molecules: $TRE(PE) < -0.01$

VI. APPLICATIONS OF THE TRE MODEL

The TRE model has been used abundantly as a criterion for aromatic stability of organic compounds[25,34,56-65] and inorganic macrocycles.[66] Here the application of the TRE model to several classes of conjugated systems will be reviewed.

A. Hückel Annulenes

Hückel [N]-annulenes have a cyclic topology which is convenient for deriving the analytical formulae for TRE. The analytical expressions for E_π of [N]-annulenes[67,68] are given below,

$$E_\pi([N]\text{-annulene}) = \begin{cases} 2\,(\text{ctan}\Theta - \tan\Theta), & N = 4m \\ 2\,\text{ctan}\Theta\,\cos\Theta, & N = 4m + 1 \text{ or } N = 4m + 3 \\ 2\,(\text{ctan}\Theta + \tan\Theta), & N = 4m + 2 \end{cases}$$

$$(32)$$

with $\Theta = \pi/2N$. The analytical expressions for the corresponding reference polynomials may be obtained by starting with the relation (89) given in Chapter 6, Volume I, for roots of linear polyenes,

$$E_j = 2 \cos \frac{j\pi}{N+1} \; ; \; j = 1,2,\ldots,N \tag{33}$$

The E_π([N]-annulene reference structure) is given as a sum of occupied E_js,

$$E_\pi([N]\text{-annulene reference structure}) = 2 \sum_{j=1}^{L} E_j =$$

$$4 \sum_{j=1}^{L} \cos \frac{j\pi}{N+1} \tag{34}$$

where

$$L = \begin{cases} \dfrac{N-2}{2} \; ; N = \text{even} \\[2ex] \dfrac{N-1}{2} \; ; N = \text{odd} \end{cases} \tag{35}$$

The Equation (34) can be further simplified using the technique of Polansky,[69]

$$E_\pi([N]\text{-annulene reference structure}) = \begin{cases} 2 \operatorname{ctg} \Theta/\sin\Theta, & N = \text{even} \\ 2 \operatorname{ctg} \Theta, & N = \text{odd} \end{cases} \tag{36}$$

The TRE expressions for [N]-annulenes are obtained by substituting (32) and (36) into (1),

$$\text{TRE}([N]\text{-annulene}) = \begin{cases} 2\left(\dfrac{\cos\Theta - 1}{\sin\Theta} - \tan\Theta\right), & N = 4m \\[2ex] 2\dfrac{\cos\Theta - 1}{\sin\Theta} \cos\Theta, & N = 4m+1 \text{ or } N = 4m+3 \\[2ex] 2\left(\dfrac{\cos\Theta - 1}{\sin\Theta} + \tan\Theta\right), & N = 4m+2 \end{cases} \tag{37}$$

The analytical expressions for TRE of annulene ions can be also derived in a similar way.[58,68] Let TRE^+, TRE^{++}, TRE^-, and TRE^{--}, respectively, stand for topological resonance energies of annulene ions $(CH)_N{}^+$, $(CH_N{}^{++}$, $(CH)_N{}^-$, and $(CH_N{}^-$, respectively, Then,

$$\text{TRE}^+ = \text{TRE} + \begin{cases} 2\sin\Theta, & N = 4m \\ -2\sin\Theta, & N = 4m+1 \\ 2\sin\Theta - 2\sin 2\Theta, & N = 4m+2 \\ 2\sin\Theta, & N = 4m+3 \end{cases} \tag{38}$$

$$\text{TRE}^{++} = \text{TRE} + \begin{cases} 4\sin\Theta, & N = 4m \\ 2\sin 2\Theta - 4\sin\Theta, & N = 4m+1 \\ 4\sin\Theta - 4\sin 2\Theta, & N = 4m+2 \\ 2(\sin\Theta + \sin 2\Theta - \sin 3\Theta), & N = 4m+3 \end{cases} \tag{39}$$

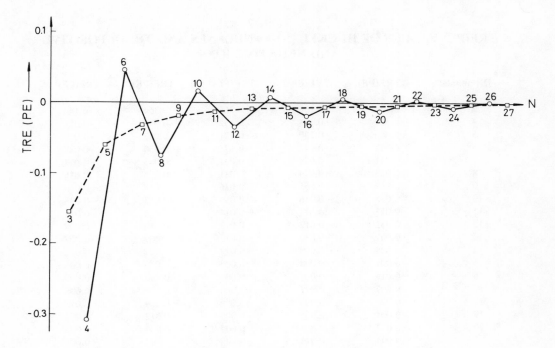

FIGURE 1. TRE(PE) vs. the ring size for [4m]-, [4m + 1]-, [4m + 2]-, and [4m + 3]-annulenes.

$$
\text{TRE}^- = \text{TRE} + \begin{cases} 2 \sin \Theta, & N = 4m \\ 2 \sin \Theta, & N = 4m + 1 \\ 2 \sin \Theta - 2 \sin 2\Theta, & N = 4m + 2 \\ -2 \sin \Theta, & N = 4m + 3 \end{cases} \tag{40}
$$

$$
\text{TRE}^{--} = \text{TRE} + \begin{cases} 4 \sin \Theta, & N = 4m \\ 2(\sin \Theta + \sin 2\Theta - \sin 3\Theta), & N = 4m + 1 \\ 4 \sin \Theta - 4 \sin 2\Theta, & N = 4m + 2 \\ 2 \sin 2\Theta - 4 \sin \Theta, & N = 4m + 3 \end{cases} \tag{41}
$$

Numerical values for TREs of [N]-annulenes and [N]-annulene ions are easily obtained from the above formulae. Predictions based on the TRE values agree nicely with the chemistry of annulenes and their ions.

The dependence of TRE(PE) on the ring size is shown in Figure 1. A plot of TRE(PE) vs. N = even indicate that both [4m]- and [4m + 2]-annulenes become rapidly nonaromatic compounds with increasing value of N and that the difference between them practically disappears at N = 26. Similarly, the odd-membered annulenes quickly converge to nonaromaticity, though the first few members appear to be antiaromatic species. Annulene positive and negative ions also converge rather rapidly to nonaromaticity (see Table 7).

B. Relationship between TREs and Ring Currents of [4m + 2] π-Electron Annulenes

According to the NMR criterion of aromaticity the diamagnetic ring currents indicate aromaticity while paramagnetic ring currents indicate antiaromaticity.[70,71] This magnetic criterion was theoretically justified when Aihara[72] showed that the diamagnetic susceptibility of a cyclic conjugated system reflects its conjugative stabilization. In the case of [N]-annulenes, the sign of susceptibility exactly agrees with that of the resonance energy.[72]

An analytical relationship between the resonance energies and reduced ring current, RCs, of [4m + 2] π-electron annulenes is derived by Haddon.[73] In analogy to this work, one

Table 7
TRE(PE) VALUES OF HÜCKEL [N]-ANNULENES AND THEIR POSITIVE
AND NEGATIVE IONS

[N]-annulene	TRE(PE)	TRE(PE)$^+$	TRE(PE)$^{++}$	TRE(PE)$^-$	TRE(PE)$^{--}$
3	−0.155	0.268	0.268	−0.366	−0.146
4	−0.307	−0.154	0.152	−0.092	0.051
5	−0.060	−0.230	−0.121	0.053	−0.018
6	0.045	−0.042	−0.173	−0.030	−0.087
7	−0.031	0.038	−0.031	−0.083	−0.027
8	−0.073	−0.029	0.031	−0.023	0.018
9	−0.019	−0.065	0.026	0.018	−0.013
10	0.016	−0.016	−0.056	−0.013	−0.038
11	−0.013	0.014	−0.014	−0.036	−0.011
12	−0.032	−0.012	0.013	−0.010	0.009
13	−0.009	−0.030	−0.011	0.009	−0.007
14	0.008	−0.008	−0.027	−0.007	−0.021
15	−0.007	0.008	−0.007	−0.020	−0.006
16	−0.019	−0.007	0.007	−0.006	0.005
17	−0.005	−0.017	−0.006	0.005	−0.006
18	0.005	−0.005	−0.016	−0.004	−0.013
19	−0.004	0.005	−0.005	−0.012	−0.004
20	−0.012	−0.004	0.004	−0.004	0.003
21	−0.004	−0.011	−0.004	0.003	−0.003
22	0.003	−0.003	−0.011	−0.003	−0.009
23	−0.003	0.003	−0.003	−0.008	−0.003
24	−0.008	−0.003	0.003	−0.003	0.002

may derive the relationship between TREs and RCs quantities of [4m + 2] π-electron annulenes. There are possibly *five* classes of [4m + 2] π-electron annulenes. These are [4m + 2]-annulenes, [4m]$^{++}$-annulenes, [4m]$^{--}$-annulenes, [4m + 1]$^-$-annulenes, and [4m + 3]$^+$-annulenes. They are all aromatic species. Topological resonance energies of these systems are given analytically in Section VI.A but here they will be transformed to more convenient forms for the further discussion,

$$\text{TRE} = \frac{4}{\sin 2\Theta} (1 - \cos \Theta), \quad [4m + 2]\text{-annulenes} \tag{42}$$

$$\text{TRE} = \frac{4}{\tan 2\Theta} (1 - \cos \Theta), \quad [4m]^+ - \text{ and } [4m]^{--}\text{-annulenes} \tag{43}$$

$$\text{TRE} = \frac{2}{\sin \Theta} (1 - \cos \Theta), \quad [4m + 1]^{--} \text{ and } [4m + 3]^+\text{-annulenes} \tag{44}$$

On the other hand, the reduced ring currents for the same annulene structures are given by,[74-76]

$$\text{RC} = \frac{2S}{N^2 \sin 2\Theta}, \quad [4m + 2]\text{-annulenes} \tag{45}$$

$$\text{RC} = \frac{2S}{N^2 \tan 2\Theta}, \quad [4m]^{++} - \text{ and } [4m]^{--}\text{-annulenes} \tag{46}$$

$$\text{RC} = \frac{S}{N^2 \sin \Theta}, \quad [4m + 1]^- - \text{ and } [4m + 3]^+\text{-annulenes} \tag{47}$$

where S is the area of the annulene ring, $\Theta = \pi/2N$, and N the size of the ring. When Equations (42) to (44) and (45) to (47) are combined, one obtains the following relation[77] for all classes of $[4m + 2]$ π-electron annulenes:

$$\text{TRE} = \frac{2N^2 RC}{S} (1 - \cos \Theta) \tag{48}$$

The expansion of the cosine function in the corresponding power series,

$$1 - \cos \Theta = 1 - \left(1 - \frac{\Theta^2}{2!} + \frac{\Theta^4}{6!} - \frac{\Theta^6}{6!} + \ldots \right) \tag{49}$$

or substituting $\Theta = \pi/2N$ into (49),

$$1 - \cos \Theta = \frac{\pi^2}{8N^2} \left(1 - \frac{\pi^2}{48 N^2} + \frac{\pi^4}{5760 N^4} - \ldots \right) \tag{50}$$

Introducing (50) into (48) and truncating the obtained relation at the first term produces the connection between TRE and RC,

$$\text{TRE} = \frac{\pi^2}{4 S} RC \tag{51}$$

This equation slightly differs from the Haddon's relation given below which relates the Dewar's resonance energies and reduced currents,

$$\text{DRE} = \frac{\pi^2}{3 S} RC \tag{52}$$

The difference is due to the fact that DRE is larger than TRE for annulenes.[68,78]

The diamagnetic susceptibility of an annulene system χ_π is given by,[79]

$$\chi_\pi = \frac{SI}{cH} \tag{53}$$

where I is the ring current intensity and H the applied magnetic field taken perpendicular to the annulene plane. The ring current intensity is defined by,

$$I = \left[8\pi^2 c \left(\frac{e^2}{hc}\right) H \right] RC \tag{54}$$

where c, e, and **h** are standard physical constants ($c = 2.997925 \times 10^8$ m sec^{-1}; $e = 1.6021917 \times 10^{-9}$ C; $h = 6.626196 \times 10^{-34}$ J (s). The combination of Equations (51), (53), and (54) leads to the relation between the topological resonance energy and the diamagnetic susceptibility,

$$\text{TRE} = \frac{1}{32 S^2} \left(\frac{h c}{e}\right)^2 \chi_\pi \tag{55}$$

Relation (55) indicates that TRE is directly proportional to the magnetic susceptibility and to the reciprocal value of the area of the annulene ring squared.

There are some results available in literature which support Equations (51) and (52). Hess et al.[80] have shown that proton chemical shifts of bisdehydroannulenes correlate well with the Hess-Schaad RE(PE) values.

C. Möbius Annulenes

The topology of Möbius annulenes coincide with that of Hückel annulenes except that the Möbius annulenes possess an edge of weight -1. The analytical expressions for TRE of Möbius [N]-annulenes are, therefore, closely related to those of Hückel [N]-annulenes and can be derived following the same reasoning. The TRE expressions for Möbius annulenes are given by,

$$\text{TRE (Möbius [N]-annulene)} = \begin{cases} 2\left(\dfrac{\cos\Theta - 1}{\sin\Theta} + \tan\right), & N = 4m \\[3mm] 2\dfrac{\cos\Theta - 1}{\sin\Theta}\cos\Theta, & N = 4m+1 \text{ or } N = 4m+3 \\[3mm] 2\left(\dfrac{\cos\Theta - 1}{\sin\Theta} - \tan\Theta\right), & N = 4m+2 \end{cases}$$

$$(56)$$

A plot of TRE(PE) vs. the ring size of Möbius annulenes (see Figure 2) reveals that the Möbius [N = even]-annulenes follow the reverse, *anti-Hückel*, rule.

Thus, Möbius [4m]-annulenes are aromatic while Möbius [4m + 2]-annulenes antiaromatic, when *m* is low. At large values of *m* both classes of even-membered Möbius annulenes converge to nonaromaticity. The behavior of odd-membered Möbius annulenes parallels the result obtained for the odd-membered Hückel annulenes. This must be so because the expressions for TRE of odd-membered Hückel and Möbius annulenes are identical.

The analytical expressions for the TRE of Möbius annulene positive and negative ions are also closely related to the corresponding expressions for Hückel annulenes,

$$\text{TRE}^+\text{(Möbius annulene)} = \text{TRE} + \begin{cases} -2\sin\Theta, & N = 4m+3 \\ 2\sin\Theta, & N = 4m+2 \\ 2\sin\Theta, & N = 4m+1 \\ 2\sin\Theta - 2\sin 2\Theta, & N = 4m \end{cases} \qquad (57)$$

$$\text{TRE}^{++}\text{(Möbius annulene)} = \text{TRE} + \begin{cases} 2\sin 2\Theta - 4\sin\Theta, & N = 4m+3 \\ 4\sin\Theta, & N = 4m+2 \\ 2(\sin\Theta + \sin 2\Theta - \sin 3\Theta), & N = 4m+1 \\ 4\sin\Theta - 4\sin 2\Theta, & N = 4m \end{cases} \qquad (58)$$

$$\text{TRE}^-\text{(Möbius annulene)} = \text{TRE} + \begin{cases} 2\sin\Theta, & N = 4m+3 \\ 2\sin\Theta, & N = 4m+2 \\ -2\sin\Theta, & N = 4m+1 \\ 2\sin\Theta - 2\sin 2\Theta, & N = 4m \end{cases} \qquad (59)$$

$$\text{TRE}^{--}\text{(Möbius annulene)} = \text{TRE} + \begin{cases} 2(\sin\Theta + \sin 2\Theta - \sin 3\Theta), & N = 4m+3 \\ 4\sin\Theta, & N = 4m+2 \\ 2\sin 2\Theta - 4\sin\Theta, & N = 4m+1 \\ 4\sin\Theta - 4\sin 2\Theta, & N = 4m \end{cases} \qquad (60)$$

Möbius annulene ions also follow an anti-Hückel rule and converge rather rapidly to nonaromaticity (see Table 8).

D. Conjugated Hydrocarbons and Heterocycles

TRE and TRE(PE) value of conjugated hydrocarbons and heterocyclics can be obtained straightforwardly by means of Equations (14) or (15). TRE(PE) values of conjugated hydrocarbons and heterocyclics parallel in most cases RE(PE) values of Hess and Schaad.[21,81,82]

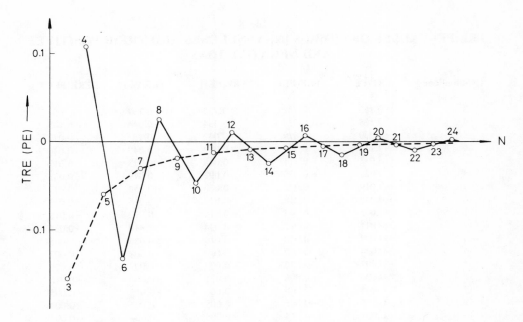

FIGURE 2. TRE(PE) vs. the ring size for the Möbius [4*m*]-, [4*m* + 1]-, [4*m* + 2]-, and [4*m* + 3]-annulenes.

A plot of TRE(PE) vs. RE(PE) for 95 randomly selected conjugated hydrocarbons and heterocyclics is given in Figure 3.

A least-squares fit produces,

$$TRE(PE) = 1.027 \, RE(PE) - 0.0065 \qquad (61)$$

with the correlation coefficient 0.977. Hence, we may conclude that both TRE(PE) and RE(PE) have practically the same predictive power for conjugated hydrocarbons and heterocyclics. However, this conclusion must be slightly modified, because it has been found[83] that the deviation of TRE(PE) from RE(PE) is proportional to the logarithm of the number of Kekulé structures K,

$$\frac{1}{a} \left\{ TRE(PE) - RE(PE) \right\} \approx \ln K \qquad (62)$$

where a is the proportionality constant. Therefore, the difference between TRE(PE) and RE(PE) is expected to be appreciable for large polycyclic molecules for which K ordinarily has large values.

E. Conjugated Ions and Radicals

The chemistry of aromatic ions and radicals has advanced in the last two decades[84] owing to improvements in preparative techniques.[85] The TRE model can be directly applied to conjugated ions and radicals. Other theoretical methods used for studying aromaticity of ions and radicals are MINDO/3 of Dewar[6] and REPA of Hess and Schaad.[24,86] REPA is the DRE-like resonance energy normalized with respect to the number of atoms in the conjugated system, i.e., REPA = resonance energy per atom.

The TRE model has been applied to numerous conjugated radicals and ions achieving in many cases a good agreement between the predictions and experimental findings.[61] The results for the pentalene radicals and ions illustrate nicely the value of the TRE approach

Table 8
TRE(PE) VALUES OF MÖBIUS [N]-ANNULENES AND THEIR POSITIVE AND NEGATIVE IONS

[N]-annulene	TRE(PE)	TRE(PE)$^+$	TRE(PE)$^{++}$	TRE(PE)$^-$	TRE(PE)$^{--}$
3	−0.155	−0.732	−0.732	0.134	0.054
4	0.108	−0.072	−0.434	−0.044	−0.145
5	−0.060	0.079	−0.042	−0.153	−0.052
6	−0.133	−0.056	0.059	−0.040	0.030
7	−0.031	−0.111	−0.048	0.028	−0.017
8	0.025	−0.025	−0.092	−0.019	−0.055
9	−0.019	0.022	−0.020	−0.052	−0.017
10	−0.047	−0.018	0.019	−0.015	0.013
11	−0.013	−0.043	−0.016	0.012	−0.010
12	0.011	−0.011	−0.038	−0.010	−0.027
13	−0.009	0.010	−0.010	−0.026	−0.008
14	−0.024	−0.009	0.009	−0.008	0.007
15	−0.007	−0.022	−0.008	0.007	−0.006
16	0.006	−0.006	−0.021	−0.006	−0.016
17	−0.005	0.006	−0.006	−0.015	−0.005
18	−0.015	−0.005	0.005	−0.005	0.004
19	−0.004	−0.014	−0.005	0.004	−0.004
20	0.004	−0.004	−0.013	−0.004	−0.011
21	−0.004	0.004	−0.004	−0.010	−0.003
22	−0.010	−0.003	0.004	−0.003	0.003
23	−0.003	−0.009	−0.003	0.003	−0.003
24	0.003	−0.003	−0.009	−0.003	−0.007

to aromatic stability. Four pentalene radicals and ions are shown below with the TRE and TRE(PE) values given underneath each molecule, TRE(PE) being given in parentheses.

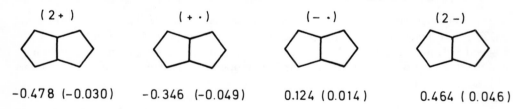

(2 +)	(+ ·)	(− ·)	(2 −)
−0.478 (−0.030)	−0.346 (−0.049)	0.124 (0.014)	0.464 (0.046)

TRE (and TRE(PE)) are consistent in their predictions suggesting pentalene cation and dication strongly antiaromatic, whereas pentalene anion and dianion aromatic. The experimental findings support theoretical results: the preparation of pentalene dication was attempted unsuccessfully,[87,88] pentalene cation is a very unstable structure,[9] substituted pentalene anion has been prepared,[88] while pentalene dianion appears to be stable[89] in contrast to the parent hydrocarbon.[8,9] The TRE and TRE(PE) values of pentalene are given below.

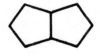

−0.216 (−0.027)

F. Homoaromatic Systems

Homoaromatic systems are nonclassical structures created in a tri-dimensional molecule when several double bonds, spatially separated, but favorably oriented, can interact through

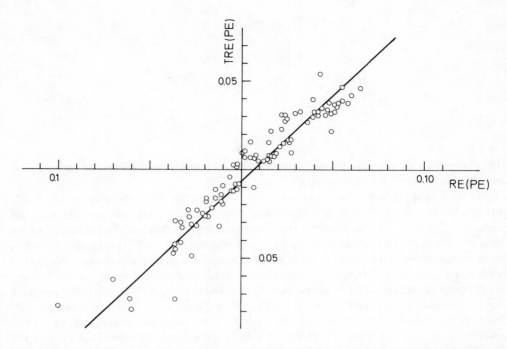

FIGURE 3.   A plot of TRE(PE) vs. RE(PE) for randomly chosen 95 conjugated molecules. The line represents a linear least-squares fit.

space to close a noose over which the π-electrons can be delocalized.[90] Hence, the formation of a polygon-like pattern by through-space interactions among disjoint intramolecular fragments is recognized as the essential feature of homoaromatic stabilization or destabilization. A good example to illustrate the homoaromatic structure is *cis, cis, cis*-cyclonona-1,4,7-triene shown below in the "crown" form.

The homoconjugation in this structure can be illustrated by the 1,3,5-trishomo hexagonal system **1**,

which in turn may be depicted by the hexagonal edge-weighted graph G_{EW}.

$$\mathsf{G_{EW}}$$

The parameter k signifies the through-space conjugative interaction between the neighboring, but, of course, not connected, oribitals.

The TRE of a homoaromatic system can be calculated by means of relation (15), while the Hückel energy of the corresponding acyclic reference structure can be obtained from Equation (21). Since the parameter k is unknown, it appears that the Hückel energy of the reference structure for homoaromatic species, and consequently its TRE value are dependent on the actual numerical value of k. Therefore, the need for a value of k is obvious. Numerical values of k have been obtained for some systems from their photoelectron spectra.[91] Thus, for example, the value for k of cyclononatriene is found to be 0.25. This value of k produces for cyclononatriene the TRE value of 0.011, which places *cis, cis, cis*-cyclonona-1,4,7-triene in the class of nonaromatic species. Therefore, the degree of conjugation in cyclononatriene is rather low (only 4% of that in a fully delocalized benzene, TRE of which is 0.276) and it perhaps best explains the nonappearance of any indication of homoaromaticity in NMR data of this compound.[92,93]

G. Aromaticity in the Lowest Excited State of Monocycles

The TRE model can be directly used for studying species in the excited state. The pioneering work in this area was carried out by Baird[94] who extended the DRE concept to the lowest $\pi\text{-}\pi^*$ triplet state of conjugated hydrocarbons by comparing the energy of triplets with the most stable biradical reference structure. Below are reported TREs of Hückel and Möbius annulenes in the lowest excited state, because this is the most interesting state for the majority of photochemical reactions.[95]

Topological resonance energy of an annulene in the excited state is denoted by TRE*. The TRE* formulae for both Hückel and Möbius [N]-annulenes are given by,

[N]-annulenes

	Hückel	Möbius

$$\text{TRE}^* = \text{TRE} + \begin{cases} 2(2\sin\Theta - \sin 4\Theta) & N = 4m & N = 4m + 2 \\ 2(\sin 2\Theta - \sin\Theta - \sin 3\Theta) & N = 4m + 1 & N = 4m + 3 \\ 4(\sin\Theta - \sin 2\Theta) & N = 4m + 2 & N = 4m \\ 2(\sin\Theta + \sin 2\Theta - \sin 5\Theta) & N = 4m + 3 & N = 4m + 1 \end{cases} \qquad (63)$$

where TRE is the topological resonance energy of the ground state while $\Theta = \pi/2N$. Numerical results for the TRE* of annulenes are reported in Table 9.

The results in Table 9 reveal several interesting points. For example, the lowest excited state of cyclobutadiene has a large and positive (aromatic) value of TRE* (0.304), while benzene has a large and negative (anti-aromatic) value of TRE* (-0.690). These results are in agreement with the work of Baird.[94] They also indicate that the Hückel 4m rings are aromatic and the 4m + 2 rings antiaromatic in the lowest excited state. However, as clearly seen from Table 9 the above is not valid for any m, but only for annulene systems with small m. With increasing m (m = even) both aromatic and antiaromatic electronically excited annulenes converge to nonaromaticity. Odd-membered annulenes in the excited state start

Table 9
TRE* AND TRE(PE)* VALUES OF
ANNULENES IN THE LOWEST
EXCITED STATE

Hückel [N]-Annulenes

[N]-annulene	TRE	TRE*	TRE(PE)*
3	−0.465	1.267	0.422
4	−1.228	0.304	0.076
5	−0.300	−1.360	−0.272
6	0.270	−0.690	−0.115
7	−0.217	−0.707	−0.101
8	−0.584	0.184	0.023
9	−0.171	−0.837	−0.093
10	0.160	−0.450	−0.045
11	−0.143	−0.605	−0.055
12	−0.384	0.144	0.012
13	−0.126	−0.644	−0.046
14	0.112	−0.336	−0.024
15	−0.105	−0.330	−0.022
16	−0.304	0.096	0.006
17	−0.085	−0.459	−0.027
18	0.090	−0.252	−0.014
19	−0.076	−0.399	−0.021
20	−0.240	0.080	0.004
21	−0.084	−0.378	−0.018
22	0.066	−0.220	−0.010
23	−0.069	−0.322	−0.014
24	−0.192	0.070	0.003

Möbius [N]-Annulenes

[N]-annulene	TRE	TRE*	TRE(PE)*
3	−0.465	−1.731	−0.577
4	0.432	−0.868	−0.217
5	−0.300	−0.510	−0.102
6	−0.798	0.234	0.039
7	−0.217	−1.043	−0.149
8	0.200	−0.552	−0.069
9	−0.171	−0.675	−0.05
10	−0.470	0.150	0.015
11	−0.143	−0.693	−0.063
12	0.132	−0.384	−0.032
13	−0.117	−0.533	−0.041
14	−0.336	0.112	0.008
15	−0.105	−0.510	−0.034
16	0.096	−0.288	−0.018
17	−0.085	−0.425	−0.025
18	−0.270	0.090	0.005
19	−0.076	−0.418	−0.022
20	0.080	−0.240	−0.012
21	−0.084	−0.357	−0.017
22	−0.220	0.066	0.003
23	−0.069	−0.345	−0.015
24	0.072	−0.192	−0.008

with very antiaromatic [3]-annulene (TRE* = -1.731) and slowly converge to nonaromaticity at higher m values.

The experimental data about the stabilities of annulenes in the excited state are rather scarce. Perhaps the high antiaromatic character of benzene in the lowest excited state best explain its reactivity in that state. For example, benzene undergoes photocycloaddition with simple olefins to produce 1,2-, 1,3-, and 1,4-adducts.[96] Similarly, benzene on irradiation with UV light forms a benzene valence isomer: Dewar benzene.[97]

TRE* values for the Möbius [N]-annulenes in the excited state point towards opposite predictions to those for Hückel annulenes: the 4m rings should be antiaromatic while the 4m + 2 rings should be aromatic for m small. Like Hückel annulenes in the lowest excited state, Möbius annulenes also converge to nonaromaticity when m increases. In the case of Möbius annulenes, experimental data are very limited indeed. However, there are proposals that photochemical reactions with the transition state having the Möbius-type structure are allowed only if the transition state has 4m + 2 topology.[98] Therefore, if the Dewar-Evans-Zimmermann rules[98,99] are accepted as valid for qualitatively determining the allowance of perycyclic reactions, then the TRE model represents a convenient measure for quantitatively treating the aromaticity of cyclic transition states.

REFERENCES

1. **Lewis, D. and Peters, D.,** *Facts and Theories of Aromaticity,* MacMillan, London, 1975, 52.
2. **Craig, D. P.,** in *Non-Benzenoid Aromatic Compounds,* Ginsburg, D., Ed., Wiley-Interscience, New York, 1959, 1.
3. **Snyder, J. P.,** in *Non-Benzenoid Aromatics,* Snyder, J. P., Ed., Academic Press, New York, 1969, 1.
4. **Badger, G. M.,** *Aromatic Character and Aromaticity,* University Press, Cambridge, 1969.
5. **Balaban, A. T.,** *Pure Appl. Chem.,* 52, 1409, 1980.
6. **Dewar, M. J. S.,** *Pure Appl. Chem.,* 44, 767, 1975.
7. **Streitwieser, A., Jr.,** *Molecular Orbital Theory for Organic Chemists,* John Wiley & Sons, New York, 1961, 239.
8. **Bergman, E. D.,** in *Non-Benzoid Aromatic Compounds,* Ginsburg, D., Ed., Wiley-Interscience, New York, 1959, 141.
9. **Lloyd, D.,** *Carbocyclic Non-Benzenoid Aromatic Compounds,* Elsevier, Amsterdam, 1966.
10. **Hall, G. G.,** *Int. J. Math. Educ. Sci. Technol.,* 4, 233, 1973.
11. **McClelland, B. J.,** *J. Chem. Phys.,* 54, 640, 1971.
12. **Trinajstić, N.,** *Croat. Chem. Acta,* 37, 307, 1965.
13. **Dewar, M. J. S.,** *The Molecular Orbital Theory of Organic Chemistry,* McGraw-Hill, New York, 1969, 176.
14. **Baird, N. C.,** *J. Chem. Educ.,* 48, 509, 1971.
15. **Dewar, M. J. S. and de Llano, C.,** *J. Am. Chem. Soc.,* 91, 789, 1969.
16. **Dewar, M. J. S. and Trinajstić, N.,** *J. Chem. Soc., A,* 1754, 1969.
17. **Dewar, M. J. S., Harget, A. J., and Trinajstić, N.,** *J. Am. Chem. Soc.,* 91, 6321, 1969.
18. **Dewar, M. J. S. and Trinajstić, N.,** *J. Am. Chem. Soc.,* 92, 1453, 1970.
19. **Murrell, J. N. and Harget, A. J.,** *Semi-empirical SCF MO Theory of Molecules,* John Wiley & Sons, London, 1972, 70.
20. **Hess, B. A., Jr. and Schaad, L. J.,** *J. Org. Chem.,* 37, 4179, 1972; see also **Gutman, I., Milun, M., and Trinajstić, N.,** *Chem Phys. Lett.,* 23, 284, 1973.
21. **Hess, B. A., Jr. and Schaad, L. J.,** *J. Am. Chem. Soc.,* 93, 305, 1971.
22. **Milun, M., Sobotka, Ž., and Trinajstić, N.,** *J. Org. Chem.,* 37, 139, 1972.
23. **Schaad, L. J. and Hess, B. A., Jr.,** *Isr. J. Chem.,* 20, 281, 1980.
24. **Hess, B. A., Jr. and Schaad, L. J.,** *Pure Appl. Chem.,* 52, 1471, 1980.
25. **Gutman, I., Milun, M., and Trinajstić, N.,** *Math. Chem. (Mülheim/Ruhr),* 1, 171, 1975.
26. **Sachs, H.,** *Publ. Math. (Debrecen),* 11, 119, 1964.
27. **Graovac, A., Gutman, I., Trinajstić, N., and Živković, T.,** *Theor. Chim. Acta,* 26, 67, 1972.
28. **Herndon, W. C. and Ellzey, M. L., Jr.,** *J. Chem. Inf. Comp. Sci.,* 19, 260, 1979.

29. **Aihara, J.-I.,** *J. Am. Chem. Soc.,* 98, 2750, 1976.
30. **Farrell, E. J.,** *J. Comb. Theory,* 26B, 111, 1979; 27B, 75, 1979.
31. **Heilmann, O. J. and Lieb, E. H.,** *Commun. Math. Phys.,* 25, 190, 1972.
32. **Hess, B. A., Jr., Schaad, L. J., and Agranat, I.,** *J. Am. Chem. Soc.,* 100, 5268, 1978.
33. **Džonova-Jerman-Blažič, B., Mohar, B., and Trinajstić, N.,** in *Applications of Information and Control Systems,* Lainiotis, D. G. and Tzannes, N. S., Eds., Reidel, Dordrecht, Holland, 1980, 395; see also **Mohar, B. and Trinajstić, N.,** *J. Comp. Chem.,* 3, 28, 1982.
34. **Gutman, I., Milun, M., and Trinajstić, N.,** *J. Am. Chem. Soc.,* 99, 1692, 1977.
35. **Trinajstić, N.,** *Int. J. Quantum Chem.,* S 11, 469, 1977.
36. **Heilbronner, E.,** *Helv. Chim. Acta,* 36, 170, 1953.
37. **Schaad, L. J., Hess, B. A., Jr., Nation, J. B., Trinajstić, N., and Gutman, I.,** *Croat. Chem. Acta,* 52, 233, 1979.
38. **Godsil, C. D. and Gutman, I.,** *Z. Naturforsch.,* 34 a, 776, 1979.
39. **Gutman, I.,** *Croat. Chem. Acta,* 54, 75, 1981.
40. **Graovac A., Kasum, D., and Trinajstić, N.,** *Croat. Chem. Acta,* 54, 91, 1981.
41. **Graovac, A.,** *Chem. Phys. Lett.,* 82, 248, 1981.
42. **Gutman, I. and Hosoya, H.,** *Theor. Chim. Acta,* 48, 279, 1978.
43. **Godsil, C. D. and Gutman, I.,** *J. Graph Theory,* 5, 137, 1981.
44. **Kurosh, A.,** *Higher Algebra,* Mir, Moscow, 1980, chap. 5.
45. **Harary, F.,** *Graph Theory,* Addison-Wesley, Reading, Mass., 1971, 96, second printing.
46. **Gutman, I.,** *Math. Chem. (Mülheim/Ruhr),* 6, 75, 1979.
47. **Gutman, I.,** *J. Chem. Soc. Faraday Trans. 2,* 799, 1979.
48. **Gutman, I. and Bosanac, S.,** *Tetrahedron,* 33, 1809, 1977.
49. **Bosanac, S. and Gutman, I.,** *Z. Naturforsch.,* 32 a, 10, 1977.
50. **Hückel, E.,** *Z. Phys.,* 76, 628, 1932.
51. **Dewar, M. J. S.,** *The Molecular Orbital Theory of Organic Chemistry,* McGraw-Hill, New York, 1969, chap. 6.
52. **Kruszewski, J. and Krygowski, T. M.,** *Can. J. Chem.,* 53, 945, 1975.
53. **Gutman, I. and Trinajstić, N.,** *Can. J. Chem.,* 54, 1789, 1976.
54. **Dewar, M. J. S. and Gleicher, G. J.,** *J. Am. Chem. Soc.,* 87, 685, 1965.
55. **Krygowski, T. M. and Kruszewski, J.,** *Quantitative Criteria of Aromaticity,* (in Polish), Wraclaw, 1978.
56. **Ilić, P., Džonova-Jerman-Blažič, B., Mohar, B., and Trinajstić, N.,** *Croat. Chem. Acta,* 52, 35, 1979.
57. **Aihara, J.-I.,** *Bull. Chem. Soc. Jpn.,* 50, 3057, 1977.
58. **Ilić, P., Sinković, B., and Trinajstić, N.,** *Isr. J. Chem.,* 20, 258, 1980.
59. **Aihara, J.-I.,** *J. Am. Chem. Soc.,* 98, 2750, 1976.
60. **Aihara, J.-I.,** *J. Am. Chem. Soc.,* 99, 2048, 1977.
61. **Ilić, P. and Trinajstić, N.,** *J. Org. Chem.,* 45, 1738, 1980.
62. **Ilić, P. and Trinajstić, N.,** *Pure Appl. Chem.,* 52, 1495, 1980.
63. **Ilić, P., Jurić, A., and Trinajstić, N.,** *Croat. Chem. Acta,* 53, 587, 1980; **Ilić, P., Mohar, B., Knop, J. V., Jurić, A., and Trinajstić, N.,** *J. Heterocycl. Chem.,* 19, 121, 1982.
64. **Ilić, P. and Trinajstić, N.,** *Croat. Chem. Acta,* 53, 591, 1980.
65. **Sabljić, A. and Trinajstić, N.,** *J. Mol. Struct.,* 49, 415, 1978; *J. Org. Chem.,* 46, 3457, 1981.
66. **Gimarc, B. M. and Trinajstić, N.,** *Inorg. Chem.,* 21, 21, 1982.
67. **Gutman, I., Milun, M., and Trinajstić, N.,** *Croat. Chem. Acta,* 44, 207, 1972.
68. **Gutman, I., Milun, M., and Trinajstić, N.,** *Croat. Chem. Acta,* 49, 441, 1977.
69. **Polansky, O. E.,** *Monat. Chem.,* 91, 916, 1960.
70. **Lewis, D. and Peters, D.,** *Facts and Theories of Aromaticity,* MacMillan, London, 1975, 25.
71. **Haddon, R. C., Haddon, V. R., and Jackman, L. M.,** *Topics Curr. Chem.,* 16, 103, 1971.
72. **Aihara, J.-I.,** *J. Am. Chem. Soc.,* 101, 558, 1979.
73. **Haddon, R. C.,** *J. Am. Chem. Soc.,* 101, 1722, 1979.
74. **London, F.,** *J. Phys. Radium,* 8, 397, 1937.
75. **McWeeny, R.,** *Mol. Phys.,* 1, 311, 1958.
76. **Haigh, C. W. and Mallion, R. B.,** in *Progress in Nuclear Magnetic Resonance Spectroscopy,* Vol. 13, Emsley, J. W., Feeney, J., and Sutcliffe, L. H., Eds., Pergamon Press, Oxford, 1979, 303.
77. **Aihara, J.-I.,** *Bull. Chem. Soc. Jpn.,* 53, 1163, 1980.
78. **Aihara, J.-I.,** *J. Am. Chem. Soc.,* 98, 6840, 1976.
79. **Salem, L.,** *Molecular Orbital Theory of Conjugated Systems,* Benjamin, New York, 1966, chap. 4.
80. **Hess, B. A., Jr., Schaad, L. J., and Nakagawa, M.,** *J. Org. Chem.,* 42, 1661, 1977.
81. **Hess, B. A., Jr., Schaad, L. J., and Holyoke, C. W., Jr.,** *Tetrahedron,* 28, 3657, 1972; 31, 295, 1975.
82. **Hess, B. A., Jr. and Schaad, L. J.,** *J. Am. Chem. Soc.,* 95, 3907, 1973.
83. **Gutman, I.,** *Bul. Chem. Soc. (Belgrade),* 43, 191, 1978; **Jashari, G., Trinajstić, N., and Velenik-Oliva, A.,** unpublished results.

84. **Isaacs, N. S.,** *Reactive Intermediates in Organic Chemistry,* Olah, G. A., Ed., John Wiley & Sons, New York, 1974.
85. **Olah, G. A., Staral, J. S., Liang, G., Paquette, L. A., Melega, W. P., and Carmody, M. J.,** *J. Am. Chem. Soc.,* 100, 3349, 1978.
86. **Bates, R. B., Hess, B. A., Jr., Ogle, C. A., and Schaad, L. J.,** *J. Am. Chem. Soc.,* 103, 5062, 1981.
87. **Trost, B. and Kinson, P. L.,** *J. Am. Chem. Soc.,* 97, 2438, 1975.
88. **Johnson, R. W.,** *J. Am. Chem. Soc.,* 99, 1461, 1977.
89. **Katz, T. J. and Rosenberg, M.,** *J. Am. Chem. Soc.,* 84, 865, 1962.
90. **Winstein, S.,** in *Aromaticity,* Spec. Publ. No. 21, The Chemical Society, London, 1967, 47.
91. **Bischof, P., Gleiter, R., and Heilbronner, E.,** *Helv. Chim. Acta,* 53, 1425, 1970.
92. **Untch, K.,** *J. Am. Chem. Soc.,* 85, 345, 1963.
93. **Untch, K., and Kurland, R. J.,** *J. Am. Chem. Soc.,* 85, 346, 1963.
94. **Baird, N. C.,** *J. Am. Chem. Soc.,* 94, 4941, 1972.
95. **Cowan, D. O. and Drisco, R. L.,** *Elements of Organic Photochemistry,* Plenum Press, New York, 1976.
96. **Wilzbach, K. E. and Kaplan, L.,** *J. Am. Chem. Soc.,* 93, 2073, 1971.
97. **Wilzbach, K. E. and Kaplan, L.,** *J. Am. Chem. Soc.,* 87, 4004, 1965.
98. **Zimmerman, H. E.,** in *Pericyclic Reactions,* Vol. 1, March, A. P. and Lehr, R. E., Eds., Academic Press, New York, 1977, 53.
99. **Dewar, M. J. S.,** *Angew. Chem. Int. Ed. Engl.,* 10, 761, 1971.

Chapter 2

ENUMERATION OF KEKULÉ STRUCTURES

I. THE ROLE OF KEKULÉ VALENCE STRUCTURES IN CHEMISTRY

Kekulé structures have been used in organic chemistry[1,2] for qualitative characterization of polycyclic conjugated systems since the last century.[3] However, until the systematic work by Pauling[4] and Wheland,[5] they have been used only occasionally.[6] The manipulations with Kekulé structures (i.e., the principal resonance structures in the qualitative valence bond theory) have shown to be useful, even nowadays, in the chemistry of conjugated molecules[7] in spite of some strong contrary opinions.[8] The use of Kekulé structures in interpreting structure and reactivity of polycyclic hydrocarbons resulted in several important achievements:

1. *The resonance theory conjecture*[5] states that the (thermodynamic) stability of isomers of conjugated structures is proportional to the number of the corresponding Kekulé structures, i.e., the most stable isomer has the greatest number of Kekulé structures,

$$\Delta K = K_2 - K_1 \implies \Delta E = E_2 - E_1 \qquad (1)$$

   where K denotes a number of Kekulé structures of a given conjugated molecule, while E stands for the energy of a molecule.

   Example

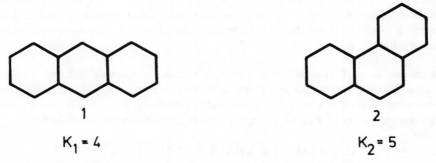

$$1 \qquad\qquad 2$$
$$K_1 = 4 \qquad\qquad K_2 = 5$$

   The stability order, determined by the Kekulé structures count, is

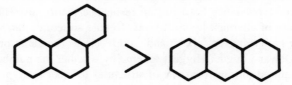

   This conjecture has been recently revised[9] and shown to hold only for benzenoid systems (see discussion later in this chapter).

2. *The Clar postulate*[10] states that a benzenoid system with no Kekulé structure (a non-Kekuléan system) should be an unstable biradical. In other words, Clar postulated nonexistence of non-Kekuléan benzenoids. Since its introduction in 1958, this postulate has remained unchallenged. Therefore, the existence of Kekulé structures is of supreme importance for the stability of aromatic systems.

   According to the number of Kekulé structures, all conjugated molecules may be classified into two groups: (1) non-Kekuléan molecules: K = 0 and (2) Kekuléan molecules: K > 0.

Example

3	4
$K_3 = 16$	$K_4 = 0$
Kekuléan molecule	Non-Kekuléan molecule

Kekuléan hydrocarbons may be partitioned into three classes: (1) polyenes: K = 1, (2) monocycles: K = 2, and (3) polycycles: K ≥ 2.

3. *The structure-resonance theory*[11] represents quantification of the resonance theory.[4,5] The essential feature of the structure-resonance theory is the restriction of the approach to the principal resonance structures, which are assumed to contribute equally to a resonance hybrid. Some of the results achieved by the structure-resonance theory will be presented in a separate section.

4. *The conjugated circuits model* of Randić[12] by which he has shown that Kekulé structures contain circuits of (4m + 2) and 4m size (m = 1,2, . . .). The contribution to the stability of a molecule is either positive or negative depending on whether the circuit is (4m + 2) or 4m. The next chapter is devoted to discussion about the conjugated circuits model.

We note that in all four mentioned results, it is rather important to know how many Kekulé structures a given conjugated molecule has. This chapter reviews the methods for the enumeration of Kekulé structures. However, before doing that we will discuss briefly two concepts: the parity of Kekulé structures and the structure-resonance theory.

II. PARITY OF KEKULÉ STRUCTURES

Dewar and Longuet-Higgins[13] introduced the concept of *parity* of Kekulé valence structures in discussing the correspondence between the resonance theory and Hückel theory. They found that in the case of alternant hydrocarbons, Kekulé structures can be separated in two classes of different parity: "even" (K^+) and "odd" (K^-). The same or different parity is determined in the Dewar-Longuet-Higgins scheme by whether the number of transpositions of double bonds required to transform one of structures to other is *even* or *odd*. If even, as in two Kekulé structures of cyclobutadiene, the valence structures are considered to be of opposite parity, while if odd, as in two Kekulé structures of benzene, the valence structures are considered to be of the same parity. Thus we get a relation "to be of the same parity" on the set of all Kekulé structures. If this relation is transitive, the set of all Kekulé structures splits up into several classes of equivalence. In the case that we have exactly two such classes (that naturally interests us most) we can denote one of them K^+ and the second K^-, so that the total number of Kekulé structures K is equal to the sum of even K^+ and odd K^- Kekulé structures,

$$K = K^+ + K^- \qquad (2)$$

Similarly, the difference between even and odd Kekulé structures is introduced as the *algebraic structure count* ASC[14],

$$\text{ASC} = \left| K^+ - K^- \right| \tag{3}$$

There are some difficulties in establishing the theoretical foundation of the parity concept. The original discussion[13] that two Kekulé structures are of the *same* parity if, and only if, one structure is obtained from the other by transposing an *odd* number of double bonds, is not applicable in all cases.[15]

Example

5

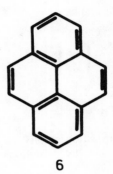

6

In spite of the fact that the structure **6** is obtained by transposition of 6 (even) double bonds of **5**, both Kekulé structures **5** and **6** of pyrene are of the same parity. This inconsistency may be removed by amending the Dewar Longuet-Higgins scheme in the following way.

Definition I: two different Kekulé structures K and K′ of the graph G are said to be of the same parity if the number of transpositions of double bonds required to transform K to K′ is odd, or if there exists a sequence of Kekulé structures of G (K_1, K_2, . . . ,K_n) such that K_1 = K and K′ = K_n and for every i = 1,2, . . . ,$n − 1$ number of transpositions of double bonds required to transform K_i to K_{i+1} is odd. Otherwise, K and K′ are said to be of the different parity.

Now it is clear that the structures **5** and **6** are of the same parity and that the above relation ''to be of the same parity'' on the set of all Kekulé structures of graph G is transitive (and, of course, reflexive and symmetrical) so that this relation is equivalence, i.e., the set of all Kekulé structures of G divides into several classes of equivalence. From chemical point of view, it is interesting when the number of these classes is exactly two, because only then we can assign them positive and negative (even and odd) parity. It is known,[15,16] that there exist graphs when this is impossible (see discussion later in this chapter).

The other way to remove the inconsistency of the Dewar-Longuet-Higgins scheme is by proposing a new definition.[15,16]

Definition II: two different Kekulé structures of G are of the same parity if the number of the $4m$-membered rings in their superposition graph is even. This definition may be analyzed in the following way. Kekulé structures and the corresponding Kekulé graphs[17] will be denoted by $a,b,c,$ . . . and $k_a,k_b,k_c,$ . . . , respectively. Let the superposition of k_a and k_b give the graph G_{ab}. Let also the number of cyclic components of the size $4m$ (m = 1,2, . . .) in the graph G be R_{4m} (G). Furthermore, let p_a = +1 if k_a is an even and p_b = −1 if k_b is an odd Kekulé structure. Then, the relation

$$p_a p_b = (-1)^{R_{4m}(G_{ab})} \tag{4}$$

determines the parity of Kekulé structures. Therefore, two Kekulé structures are of the same (opposite) parity if, and only if, $R_{4m}(G_{ab})$ is even (odd).

Example

Parity of Kekulé structures of benzocyclobutadiene
   (i) Kekulé structures of benzocyclobutadiene

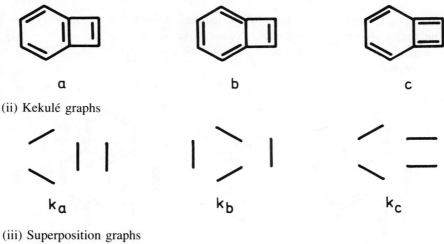

   a   b   c

   (ii) Kekulé graphs

   k_a k_b k_c

   (iii) Superposition graphs
        $p_a = +1$ (by definition)

   $G_{ab} =$ ⬡ │   ⟹ $p_b = +1$

   $G_{ac} =$ ▱   ⟹ $p_c = -1$

   (iv) Structure count

$$K^+ = 2$$
$$K^- = 1$$
$$K = K^+ + K^- = 3$$
$$ASC = \left| K^+ - K^- \right| = 1$$

A disadvantage of the definition II is that it is not clear if in this case the relation "to be of the same parity" is transitive, although so far such an example has not been found.

A. The Breakdown of the Parity Concept

The expression (4) can uniquely determine the relative parity of any *two* Kekulé structures of a molecule including nonbenzenoid and nonalternant systems. For molecules with *three or more* Kekulé structures a question of compatibility of such pairwise assignment may be raised.[15,16] If *a, b* and *c* denote three Kekulé structures of an arbitrary molecule the following situations may arise:

	Case I	Case II	Case III	Case IV
Pairs of the same parity	a,b	a,b	a,b	
	a,c	a,c		
	b,c			
Pairs of the opposite parity		b,c	a,c	a,b
			b,c	a,c
				b,c

Only Case I and Case III are internally consistent and correspond to the situation when all three structures are of the same parity and the situation when one structure is of the opposite parity, respectively. In Cases II and IV there is a contradiction. Note that in the case (Cases I and III) when we have two classes of equivalence (assigned as a class positive parity and a class of negative parity) the product of expression (4) is *positive*, otherwise (Cases II and IV) is *negative*. Therefore, a product

$$(-1)^{R_{4m}(G_{ab}) + R_{4m}(G_{ac}) + R_{4m}(G_{bc})} \tag{5}$$

indicates whether three structures a, b and c have a logically coherent assignment. When its value is $+1$ we can separate the three structures into classes of the same or opposite parity, but this is no longer possible when product is -1.

$$(-1)^{R_{4m}(G_{ab}) + R_{4m}(G_{ac}) + R_{4m}(G_{bc})} = \begin{cases} +1 \text{ Kekulé-separable} \\ \quad\text{structures} \\ \\ -1 \text{ Kekulé-insepar-} \\ \quad\text{able structures} \end{cases} \tag{6}$$

Alternatively the sum,

$$R_{4m}(abc) = R_{4m}(G_{ab}) + R_{4m}(G_{ac}) + R_{4m}(G_{bc}) \tag{7}$$

which can be either *even* or *odd*, determines the character of the case. When $R_{4m}(abc)$ is *odd*, it is not possible to assign the parity type to individual structures without contradicting already made assignments.

$$R_{4m}(abc) = \begin{cases} \text{even} \quad \text{Kekulé-separable} \\ \qquad\quad \text{structures} \\ \\ \text{odd} \quad \text{Kekulé-inseparable} \\ \qquad\quad \text{structures} \end{cases} \tag{8}$$

Generally, for systems with more than three Kekulé structures the same reasoning applies to any triplet of structures, and if there is a single such a triplet with associated $R_{4m}(xyz)$ which is *odd* we confront the case of nonseparability of Kekulé structures in two classes of opposite parity.

In order to see if conjugated systems can introduce the situations when the nonseparability of Kekulé structures occurs, a systematic examination of topologies of such molecules which have three or more Kekulé structures needs to be undertaken. Benzenoid systems contain only rings of size $4m + 2$, hence all structures consisting of hexagonal building blocks are of the same parity and no problem arises. Nonbenzenoid systems such as benzocyclobutadiene possess structures of opposite parity, but do not introduce incompatibilities. Two fused rings of $4m$ and $4m' + 2$ size, as in benzoderivatives of cyclobutadiene, produce upon the superposition rings of conjugation of size $4m$, $4m' + 2$, and $4(m + m')$, which is consistent with two structures of the same parity and one of the opposite parity. When $4m$ and $4m + 2$ rings are not fused again the structures may be separated into two classes of opposite parity. The same is also true when both rings considered are of $4m$ and $4m'$ size.

A possible source for parity incompatibility are the nonalternant systems. Bicyclic non-alternants such as azulene or pentalene, have only two Kekulé structures. Tricyclic nonal-ternant systems with condensed rings may introduce a novel situation. However, if one of the rings in the tricyclic nonalternate structure is even-membered, such a system cannot generate parity incompatibility.

Example

(i) Kekulé structures

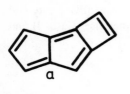 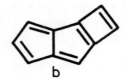 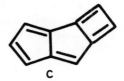

a   b   c

(ii) Kekulé graphs

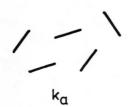  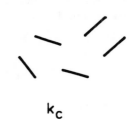

k_a k_b k_c

(iii) Superposition graphs

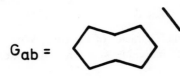

$G_{ab} =$ $R_{4m}(G_{ab}) = 1$

$G_{ac} =$ $R_{4m}(G_{ac}) = 0$

$G_{bc} =$ $R_{4m}(G_{bc}) = 1$

(iv) Structure count

$$R_{4m}(abc) = even$$

$$p_a = +1 \text{ (by definition)}$$

$$p_b = -1$$

$$p_c = +1$$

$$K^+ = 2; K^- = 1$$

$$K = 3; ASC = 1$$

Analysis of a tricyclic system with fused odd-membered rings only established that this is a system for which Kekulé structures cannot be separated in even and odd.

Example

acepentylene

(i) Kekulé structures

a

b

c

(ii) Kekulé graphs

k_a

k_b

k_c

(iii) Superposition graphs

G_{ab}

G_{ac}

G_{bc}

$$R_{4m}(G_{ab}) = 1 \qquad R_{4m}(G_{bc}) = 1 \qquad R_{4m}(G_{ac}) = 1$$

(iv) Structure count

$$R_{4m}(abc) = odd$$

$$p_a = +1 \;\; (\text{by definition})$$

$$G_{ab} \Longrightarrow p_b = -1$$

$$G_{ac} \Longrightarrow p_c = -1; \quad G_{bc} \quad p_c = +1$$

Therefore, Kekulé structures of acepentylene cannot be separated in two classes.
Systems with Kekulé-inseparable structures are expected to exhibit low stability and to

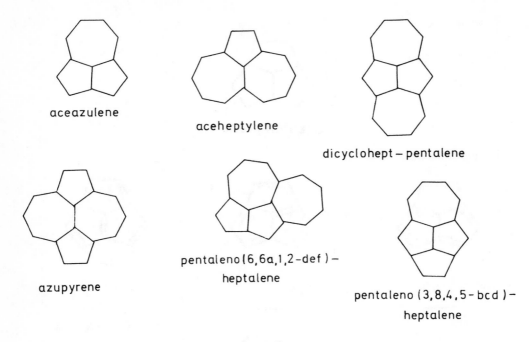

aceazulene

aceheptylene

dicyclohept – pentalene

azupyrene

pentaleno (6,6a,1,2-def) – heptalene

pentaleno (3,8,4,5- bcd) – heptalene

FIGURE 1.   Several Kekulé-inseparable systems.

be rather reactive compounds. For example, the derivative of the acepentylene system is reported[18] as dilithium hexachloroacepentylendiide, a stable species only at low temperature (230 K). In Figure 1, additional molecular systems for which the parity concept breaks down are given.

B. The Analysis of the Resonance Theory Conjecture

The determinant of the adjacency matrix **A** is related to the algebraic structure count[14,19]

$$\det \mathbf{A} = (-1)^{N/2} (K^+ - K^-)^2 \tag{9}$$

In addition, the determinant of **A** is equal to product of the eigenvalues of $\mathbf{A}$[13],

$$\det \mathbf{A} = \prod_{i=1}^{N} \left| x_i \right| \tag{10}$$

Equations (9) and (10) may be combined and expressed as

$$(ASC)^2 = \prod_{i=1}^{N} \left| x_i \right| \tag{11}$$

and further as,

$$2 \ln (ASC) = \sum_{i=1}^{N} \ln \left| x_i \right| \tag{12}$$

In the presence of NBMOs because $(ASC)^2 = 0$, Equation (11) should be altered to,[9]

$$(asc)^2 = \prod_{i=1}^{N}{}' \left| x_i \right| \tag{13}$$

where Π' indicates multiplication only over the nonzero elements of the graph spectrum. Since the $|x_i s|$ lie in the interval $(0,3)$,[20] $\ln|x|$ may be approximated by a finite polynomial,

$$P_m(x) = \sum_{t=0}^{m} (-1)^{t+1} p_t \, x^t \qquad (14)$$

Necessary conditions for $P_m(x)$ to be a good approximation to $\ln|x|$ are $p_t > 0$ and $p_1 < p_2 < \ldots < p_m$.[21] Substitution of $P_m(x)$ into Equation (12) yields

$$2 \ln (ASC) = \sum_t (-1)^{t+1} p_t A_t \qquad (15)$$

where

$$A_t = \sum_{i=1}^{N} |x_i|^t \qquad (16)$$

The advantage of introducing Equation (16) is that A_ts may be easily related to molecular topology. Thus, it is known that[22-24]

$$A_0 = N \qquad (17)$$

$$A_1 = E_\pi \qquad (18)$$

$$A_2 = 2M \qquad (19)$$

$$\cdots \cdots$$

where N and M are the number of atoms and bonds, respectively, in a molecule. Substituting Equations (17) to (19) into (15) and neglecting higher order terms the three-parameter topological formula for E_π is obtained,

$$E_\pi = (p_0/p_1)N + (2p_2/p_1)M + (2/p_1) \ln (ASC) \qquad (20)$$

or, if (p_0/p_1), $(2p_2/p_1)$, and $(2/p_1)$ are replaced by constants a, b, and c, respectively,

$$E_\pi = aN + bM + c \ln (ASC) \qquad (21)$$

The precision of this rather simple formula for approximating the Hückel energy (the values of the coefficients $a = 0.713$, $b = 0.347$, and $c = 0.765$ are obtained by a least-squares fitting to an arbitrarily selected group of conjugated hydrocarbons) is good, with an average error of 2% in E_π.[9] However, the important aspect of this formula is not so much the accuracy of predicting E_π, but what one can learn from it. Equation (21) offers a simple explanation that for a set of structural isomers (having the same number of atoms, bonds, and the same ring size, and the same type of skeleton branching) E_π is linearly increasing with $\ln(ASC)$. Thus, the important conclusion follows that values of ASC may be used as a sufficient criterion for predicting relative thermodynamic stabilities (since E_π is related to the thermodynamic parameter of a conjugated system[25]) of isomers.

Example

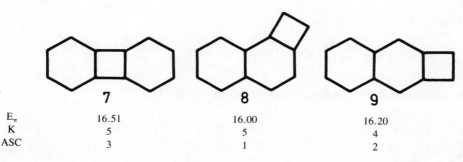

	7	8	9
E_π	16.51	16.00	16.20
K	5	5	4
ASC	3	1	2

The predicted stability order based on ASC values: **7** > **9** > **8** is in agreement with experimental findings and TRE calculations (TRE [7] = 0.120, TRE [8] = −0.384, TRE [9] = −0.120)[26] and contradicts the prediction of the traditional resonance theory: 7 ≈ 8 > 9, based on the total number of Kekulé structures.

The early successful application of traditional resonance theory using the total number of Kekulé structures was based on the fortunate fact that for benzenoid hydrocarbons and acyclic polyenes K = ASC. However, K ≠ ASC for nonbenzenoid compounds and the correction is essential there.

Finally, it should be emphasized here that the dependence of E_π on ASC is logarithmic, so that ASC gives only a small, second order, contribution to E_π. This is, of course, in agreement with the fact that E_π is mainly determined by N and M (see discussion in Chapter 6, Volume I, Section X.C).

III. ELEMENTS OF STRUCTURE-RESONANCE THEORY

Structure-resonance theory is essentially a quantification of the qualitative valence-bond resonance theory. The fundamental postulate of the structure-resonance theory is that only Kekulé structures are needed for a description of the conjugated system and that they all contribute with equal weight.[7,11,27-29] The equal weight postulate may be symbolized as

$$|\Psi> = K^{-1/2} \sum_{j=1}^{K} |j>$$ (22)

where $|\Psi>$ describes the conjugated molecule, while $|j>$ stands for the j-th Kekulé structure in a Kekulé structure basis $|1>, |2>, \ldots, |K>$. This basis is orthonormal,

$$<i|j> = \delta_{ij} \begin{cases} 1 & i = j \\ 0 & i \neq j \end{cases}$$ (23)

where $<i|j>$ denotes a matrix element between the Kekulé structures i and j.

Equation (22) is valid only for benzenoid systems. However, if we wish to include in our considerations cyclobutadienoids then it needs to contain the parity term. An improvement of Equation (22) along these lines is as follows,

$$|\Psi> = K^{-1/2} \sum_{j=1}^{K} P_j |j>$$ (24)

where P_j is the parity of the j-th Kekulé structures. Since for benzenoid systems all P_j = +1, Equation (24) immediately reduces to (22).

The total π-electron energy can be formally obtained from $|\Psi>$ by the action of some (undefined) energy operator $\hat{H}$,[30]

$$E_\pi = <\Psi|\hat{H}|\Psi>$$ (25)

Introducing (24) into (25) gives,

$$E_\pi = K^{-1} \sum_{i=1}^{K} \sum_{j=1}^{K} P_i P_j <i|H|j>$$ (26)

The resonance energy of a conjugated molecule is defined by,[11]

$$RE = E_\pi - <i|\hat{H}|i>$$ (27)

where all matrix elements $<i|\hat{H}|i>$ are the same and equal to the energy of an isolated Kekulé structure. Then,

$$RE = \frac{2}{K} \sum_{i<j} P_i P_j < i \left| \hat{H} \right| j > \tag{28}$$

or

$$RE = \frac{2}{K} \sum_{i<j} P_i P_j H_{ij} \tag{29}$$

where

$$H_{ij} = < i \left| \hat{H} \right| j > \tag{30}$$

For benzenoids, Equation (29) reduces to

$$RE = \frac{2}{K} \sum_{i<j} H_{ij} \tag{31}$$

Resonance between the Kekulé structures gives rise to several integrals.[30] However, only two are considered by Herndon[11]: γ_1 representing the resonance between two Kekulé structures related by a permutation of *three* double bonds within a six-membered ring and γ_2 which involves the permutation of *five* double bonds within a ten-membered ring or within two annelated rings. Thus, Equation (31) can be transformed to

$$RE = \frac{2}{K} (n_1 \gamma_1 + n_2 \gamma_2) \tag{32}$$

where n_1 and n_2 are the numbers (count) of each type of integral γ_1 and γ_2. The optimal values of γ_1 (= 0.841 eV) and γ_2 (= 0.336eV) are obtained[11] by comparing REs and DREs of Dewar and de Llano.[31] Introduction of the numerical values of γ_1 and γ_2 to (32) leads to the expression for the practical calculation of resonance energies,

$$RE(eV) = \frac{2}{K} (0.841 \, n_1 + 0.336 \, n_2) \tag{33}$$

which gives an average difference between REs and DREs of ± 0.041 eV.

Example

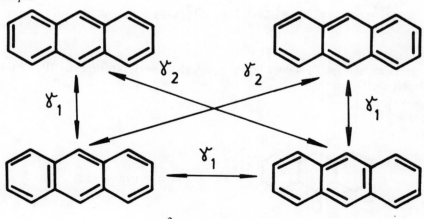

$$RE = \frac{2}{4} (3\gamma_1 + 2\gamma_2) = 1.60 \text{ eV}$$

$$DRE \text{ (Dewar } - \text{ de Llano)} = 1.60 \text{ eV}$$

There is no difficulty in parametrizing Equation (29) to cover benzenoids and cyclobutadienoids. Besides γ_1 and γ_2 already determined two additional parameters are needed: ω_1 representing the resonance between two structures related by permutation of *two* bonds within a four-membered ring and ω_2 which involves the permutation of *four* bonds within an eight-membered ring or within two annelated rings with eight-membered perimeter. Thus, Equation (29) in order to include both benzenoid and cyclobutadienoid systems, transforms to

$$RE = \frac{2}{K}(n_1\gamma_1 + n_2\gamma_2 + m_1\omega_1 + m_2\omega_2) \qquad (34)$$

where the symbols have their previous meaning, while m_1 and m_2 are the numbers (count) of each type of integral ω_1 and ω_2. The best values ω_1 ($= -0.650$ eV) and ω_2 ($= -0.260$ eV) are again found by comparing REs and DREs obtained from SCF calculations.[31] Thus, the practical formula for calculating resonance energies is given by,

$$RE = \frac{2}{K}(0.841\, n_1 + 0.336\, n_2 - 0.650\, m_1 - 0.260\, m_2) \qquad (35)$$

Equation (35) has been successfully used for calculation of resonance energies of various polycyclic conjugated hydrocarbons.[11]

Herndon[32] has found an interesting relationship between the first (adiabatic) ionization potential IP of the conjugated hydrocarbon G, and its resonance energy RE (G) and the resonance energy of the corresponding radical cation RE (G^+),

$$IP = a + RE(G) - REG^{\cdot\,+}) \qquad (36)$$

The resonance energies of radical cations and parent hydrocarbons have been calculated by means of the logarithmic functions of ASC,

$$RE(G) = b\ln[ASC(G)] \qquad (37)$$

$$RE(G^{\cdot\,+}) = c\ln[ASC(G^{\cdot\,+})] \qquad (38)$$

The least-squares fit of Equation (37) to the Dewar-de Llano DRE values produced $b = 1.185$ eV.[33] Experimental IP of 29 compounds were used to obtain the numerical values[32] of a ($= 11.277$ eV) and c ($= 1.044$ eV) and to establish the following linear relationship,

$$IP(eV) = 11.277 + 1.185\ln[ASC(G)] - 1.044\ln[ASC(G^{\cdot\,+})] \qquad (39)$$

which gives an average deviation between calculated and experimental IPs of 0.016 eV (correlation coefficient 0.995).[32]

Example

G

$SC(G) = K(G) = 4$

$RE(G) = 1.185\ln K(G)$

$= 1.185\ln 4 = 1.643$ eV

$ASC(G^{\cdot\,+}) = 2\,[SC(G)\, n_{C=C}(G) + D(G)]$

This relation was proposed by Randić.[34]
(Note, D(G) is the number of Dewar structures of a hydrocarbon)

$$D(G) = 48 \text{ (see later a way to enumerate the number of Dewar structures)}$$

$$ASC(G^{\cdot+}) = 2[4 \cdot 7 + 48] = 152$$

$$RE(G^{\cdot+}) = 1.044 \ln ASC(G^{\cdot+})$$

$$= 1.044 \ln 152 = 5.245 \text{ eV}$$

$$IP \text{ (theoretical)} = 11.227 + 1.643 - 5.245 = 7.68 \text{ eV}$$

$$IP \text{ (experimental)} = 7.43 \text{ eV}$$

IV. ENUMERATION OF RESONANCE STRUCTURES

The question of enumeration of all Kekulé structures has been continuously discussed in literature.[7,35-40] This is so because the structure count (K and ASC) information is always in qualitative agreement with experimental manifestations of stability and reactivity. Here we will summarize several procedures for enumerating the Kekulé structures. All these enumeration schemes may be divided into several groups according to the underlying principle involved. These include enumeration via the different polynomial representations, enumeration based on the lattice structure of fused benzenoids, enumeration utilizing the properties of the coefficients of the NBMOs, enumeration based on the graph decomposition, etc. However, several techniques have been already mentioned in the preceding pages of this chapter, e.g., the number of Kekulé structures of benzenoid hydrocarbons is equal to square root of the product of the Hückel eigenvalues,[13]

$$K^2 = \prod_{i=1}^{N} |x_i| \tag{40}$$

V. ENUMERATIONS BASED ON POLYNOMIALS

A. The Characteristic Polynomial

The absolute value of the last coefficient of the characteristic polynomial of a benzenoid graph G, $|a_N(G)|$, is related to the number of Kekulé structures of a corresponding conjugated molecule,

$$\left| a_N(G) \right| = K^2 \tag{41}$$

Example

G

Let us construct a_n of anthracene graph by counting all Sachs graphs.

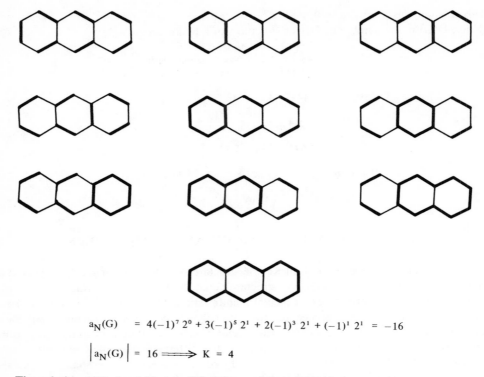

$$a_N(G) = 4(-1)^7 2^0 + 3(-1)^5 2^1 + 2(-1)^3 2^1 + (-1)^1 2^1 = -16$$

$$\left| a_N(G) \right| = 16 \implies K = 4$$

The relation (41) should be modified for cyclobutadienoid structures as

$$\left| a_N(G) \right| = (ASC)^2 \tag{42}$$

This procedure is a general one, but rather unwieldy because of the tedium of counting Sachs graphs. Beyond a certain size of molecule, it is almost impossible to carry out because of the enormous increase in the number of Sachs graphs. However, there is another path open for construction of the characteristic polynomial, that is the expansion of the adjacency matrix **A** in the form of a secular determinant:

$$A \rightarrow \det \left| xI - A \right| \rightarrow \sum_{n=0}^{N} a_n x^{N-n}$$

The procedure may terminate at this point or one may continue to obtain the roots of P(G; x). Then the positive roots of the characteristic polynomial N_+ may be also used for enumerating the Kekulé structures of benzenoids, i.e.,

$$K = \prod_{j=1}^{N_+} x_j \tag{43a}$$

or ASC of cyclobutadienoids,

$$ASC = \prod_{j=1}^{N_+} x_j \tag{43b}$$

This procedure is very practical because it can be easily adopted for a computer and, therefore, it is recommended.

Example

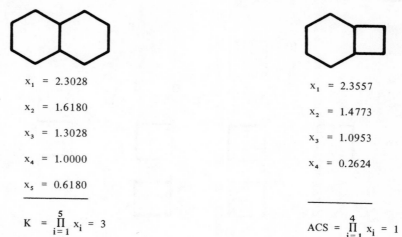

$x_1 = 2.3028$

$x_2 = 1.6180$

$x_3 = 1.3028$

$x_4 = 1.0000$

$x_5 = 0.6180$

$$K = \prod_{i=1}^{5} x_i = 3$$

$x_1 = 2.3557$

$x_2 = 1.4773$

$x_3 = 1.0953$

$x_4 = 0.2624$

$$ACS = \prod_{i=1}^{4} x_i = 1$$

B. The Acyclic Polynomial

The absolute value of the a_N^{ac} (G) coefficient of the acyclic polynomial is equal to the number of Kekulé structures for both benzenoid and cyclobutadienoid systems,

$$\left| a_N^{ac}(G) \right| = K \tag{44}$$

This result appears to be very convenient for enumeration of Kekulé structures for molecules of moderate size, but with increasing size the method becomes rather impractical.

Example

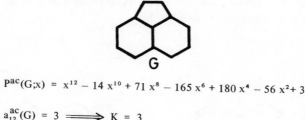

$$P^{ac}(G;x) = x^{12} - 14\,x^{10} + 71\,x^8 - 165\,x^6 + 180\,x^4 - 56\,x^2 + 3$$

$$a_{12}^{ac}(G) = 3 \implies K = 3$$

C. The Hosoya Polynomial

Hosoya[41] has introduced a polynomial, called the Z-polynomial, which is similar to the acyclic polynomial. The Hosoya polynomial or the Z-counting polynomial of a graph is defined as

$$P^h(G;x) = \sum_{k=0}^{[N/2]} p(G;k)\,x^k \tag{45}$$

where the numbers $p(G;k)$ represent the number of ways in which k disconnected K_2 components can be imbedded into G as a subgraphs.

Note, $p(G;O) = 1$ by definition. Some $p(G;k)$ values have interesting features, i.e.,

$$p(G;1) = \text{the number of edges in G} \tag{46}$$

$$p\left(G;\frac{N}{2}\right) = K \tag{47}$$

This last result serves for counting Kekulé structures.

Example

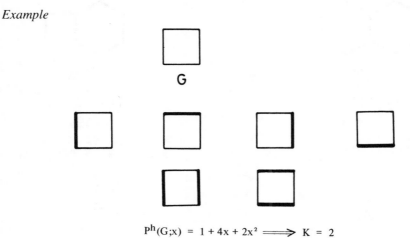

$$P^h(G;x) = 1 + 4x + 2x^2 \implies K = 2$$

The sum of the p(G; k) numbers is equal to the Z-index of a structure. Thus, the Z-index of cyclobutadiene is 7.

D. The Sextet Polynomial

The sextet polynomial has been introduced by Hosoya and Yamaguchi[42] as a convenient device for the enumeration of Clar's sextets,[43] a concept which originate from the early work by Armitt and Robinson.[44] In a given Kekulé structure, an aromatic sextet is defined as a set of three double bonds circularly conjugated as in either of the two Kekulé structures of benzene and is represented by a circle. Two (three, etc.) rings are mutually resonant if there exists a Kekulé structure such that all of these rings possess an aromatic sextet, provided that no two sextets have common double bonds.

Example

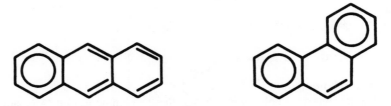

The empirical fact[45] that the angular annellation (as in phenanthrene) yields extra stabilization energy relative to the linear annellation (as in anthracene) is translated into the comparison of the maximal numbers of aromatic sextets.

The sextet polynomial is defined as

$$P^s(G;x) = \sum_{k=0}^{m} r(G;k) x^k \qquad (48)$$

where r(G; k) is the resonant sextet number of the benzenoid graph G, while m is the largest number of mutually resonant sextets.

The sextet number r(G; k) of G is defined as the number of ways in which k mutually resonant sextets are chosen from G. Note, r(G; O) = 1 by definition.

The sextet polynomial has a number of interesting properties. We list some of them below:

1. The sextet polynomial can be used for enumeration of Kekulé structures of *catafusenes*, i.e., polycyclic aromatic hydrocarbons in which no three hexagons have a common carbon atom,

$$P^s(G; x=1) = K(G) \qquad (49)$$

It could be also used to enumerate Kekulé structures of *perifusenes* which do not contain the coronene skeleton. If a perifusene contains coronene a correction must be introduced.[42] The sextet polynomial is corrected in this case by introducing the *super-ring*, i.e., a ring composed of a six hexagons forming the peripheral structure of coronene. A super-ring correction gives an additional term to x.

2. The derivative of $P^s(G; x)$ with respect to x, $p^{s'}(G; x = 1)$ is equal to the number of Kekulé structures of the subgraphs G-b_i, obtained by deleting the benzene rings b_i and adjacent bonds from G.

$$P^{s'}(G; x=1) = \sum_i K(G - b_i) \qquad (50)$$

3. The ratio of $K(G$-$b_i)$ and $K(G)$ is a measure of the aromaticity of the individual six-membered rings and is identical to the *local aromaticity index*, LAI, proposed by Randić,[46]

$$(LAI)_i = \frac{2K(G - b_i)}{K(G)} \qquad (51)$$

4. The ratio of

$$\sum_i K(G - b_i) \text{ and } K(G), \text{ or } P^{s'}(G; x=1) \text{ and } P^s(G; x=1)$$

is a measure of the *total molecular aromaticity index*, TAI, also introduced by Randić,[46]

$$TAI = \sum_i (LAI)_i = \frac{2\sum_i K(G - b_i)}{K(b)} = \frac{2P^{s'}(G; x=1)}{P^s(G; x=1)} \qquad (52)$$

Example

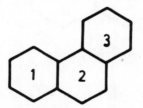

(a) Construction of the sextet polynomial r(G; O) = 1 (by definition)

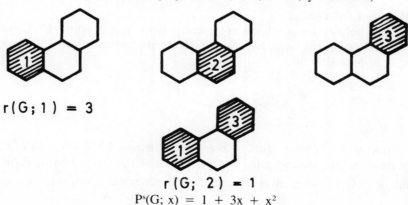

r(G; 1) = 3

r (G; 2) = 1

$$P^s(G; x) = 1 + 3x + x^2$$

(b) Enumeration of Kekulé structures

$P^s(G; x = 1) = K(G) = 5$

(c) Enumeration of Kekulé structures of subgraphs G-b_i

(c.1) $P^{s'}(G; x) = 3 + 2x$

$$P^{s'}(G; x = 1) = \sum_i K(G - b_i) = 5$$

(c.2)

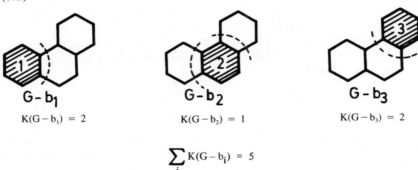

$$G - b_1 \qquad\qquad G - b_2 \qquad\qquad G - b_3$$

$$K(G - b_1) = 2 \qquad K(G - b_2) = 1 \qquad K(G - b_3) = 2$$

$$\sum_i K(G - b_i) = 5$$

(d) Evaluation of the local aromaticity index

$$(LAI)_1 = \frac{4}{5}; \quad (LAI)_2 = \frac{2}{5}; \quad (LAI)_3 = \frac{4}{5}$$

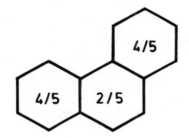

(e) Evaluation of the total aromaticity index

$$TAI = (LAI)_1 + (LAI)_2 + (LAI)_3 = 2$$

E. The Wheland Polynomial

The Wheland polynomial of a linear polyene or a benzenoid structure, $P^W(G; x)$, enumerates not only Kekulé structures, but also the other canonical structures of each degree of excitation. It is defined[47-49] as,

$$P^W(G; x) = \sum_{k=0}^{L} W(G; k)\, x^k \qquad\qquad (53)$$

where W(G; k) is the number of the k-th excited structures for a given set of canonical resonance structures of the conjugated molecule. Note, W(G; O) = 1 by definition.

The sum of W(G; k) values gives the total number of resonance structures of a molecule,

$$P^W(G;x=1) = \sum_{k=0}^{L} W(G;k) = \frac{(2n)!}{n!(n+1)!}$$ (54)

where n is the number of double bonds in a molecule.

The coefficients of the Wheland polynomial may be obtained by using the graph-theoretical rules which depend on whether one is interested in linear polyenes or benzenoid structures.

Linear polyenes

The Wheland polynomials for linear polyenes can be generated by the recurrence formula,

$$P^W(L_n;x) = p^W(L_{n-1};x) + x \sum_{k=2}^{n} P^W(L_{k-1};x) P^W(L_{n-k};x)$$ (55)

where the symbol L_n stands for linear structures with n ($= N/2$) double bonds. If one adopts $P^W(L_o; x) = 1$ and $P^W(L_1; x) = 1$, then the Wheland polynomials for higher linear polyenes may be easily obtained.

Example

$$L_4 \qquad\qquad L_3$$

$$P^W(L_4;x) = P^W(L_3;x) + x \sum_{k=2}^{4} P^W(L_{k-1};x) P^W(L_{n-k};x)$$

$$= P^W(L_3;x) + x\left\{ P^W(L_1;x) P^W(L_2;x) + P^W(L_2;x) \right.$$

$$\left. P^W(L_1;x) + P^W(L_3;x) P^W(L_0) \right\} =$$

$$1 + 6x + 6x^2 + x^3$$

$$P^W(L_4;x=1) = 14$$

All resonance structures of octatetraene, C_8H_{10} are depicted in Figure 2.

In Table 1, several Wheland polynomials of linear polyenes are given.

The inspection of coefficients of Wheland polynomials in Table 1 reveals a regularity which leads to the general expression for coefficients,

$$P^W(L_n;x) = \frac{1}{n} \sum_{k=0}^{n-1} \binom{n}{k}\binom{n}{k+1} x^k$$ (56)

Benzenoid structures

The Wheland polynomials for benzenoid structures may be constructed by using the following recurrence formula,

$$P^W(G;x) = P^W(G-e;x) + (1-x)P^W[G-(e);x]$$ (57)

where G is a graph of a given benzenoid molecule, G-e denotes a subgraph obtained by removing the edge e from G, while G-(e) represents a subgraph obtained by deletion of the

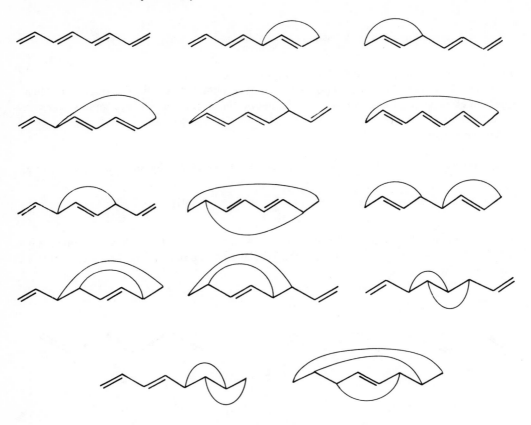

FIGURE 2.   Canonical resonance structures of octatetraene.

edge *e* and the incident vertices from G. The guiding conception behind relation (57) is to carry out the decomposition of G until it is reduced to components corresponding only to polyenes for which the Wheland polynomials are readily available.

Example

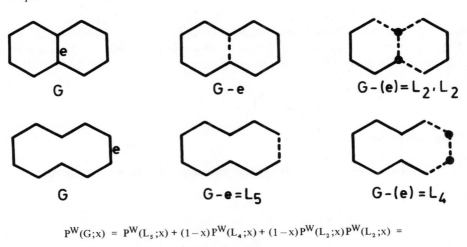

$$P^W(G;x) \;=\; P^W(L_5;x) + (1-x)P^W(L_4;x) + (1-x)P^W(L_2;x)P^W(L_2;x) \;=$$

$$3 + 16x + 19x^2 + 4x^3$$

Table 1
WHELAND POLYNOMIALS OF SOME LINEAR POLYENES

		$P^w(L_n; x = 1)$
$P^w(L_2; x) =$	$1 + x$	
$P^w(L_3; x) =$	$1 + 3x + x^2$	2
$P^w(L_4; x) =$	$1 + 6x + 6x^2 + x^3$	5
$P^w(L_5; x) =$	$1 + 10x + 20x^2 + 10x^3 + x^4$	14
$P^w(L_6; x) =$	$1 + 15x + 50x^2 + 50x^3 + 15x^4 + x^5$	42
$P^w(L_7; x) =$	$1 + 21x + 105x^2 + 175x^3 + 105x^4 + 21x^5 + x^6$	132
$P^w(L_8; x) =$	$1 + 28x + 196x^2 + 490x^3 + 490x^4 + 196x^5 + 28x^6 + x^7$	429
$P^w(L_9; x) =$	$1 + 36x + 336x^2 + 1176x^3 + 1764x^4 + 1176x^5 + 336x^6 + 36x^7 + x^8$	1430
$P^w(L_{10}; x) =$	$1 + 45x + 540x^2 + 2520x^3 + 5292x^4 + 5292x^5 + 2520x^6 + 540x^7$ $+ 45x^8 + x^9$	4862
		16796

The inspection of coefficients of Wheland polynomials reveals the following relations,

$$W(G;0) = K \tag{58}$$

$$W(G;1) = D \tag{59}$$

$$W(G;2) = D_2 \tag{60}$$

$$\dots\dots\dots\dots$$

$$W(G;L) = D_L \tag{61}$$

where symbols K, D, D_2, . . . , D_L stands for Kekulé structures, Dewar structures, doubly excited Dewar structures, . . . , L-tuple excited Dewar structures, respectively.

The above Wheland polynomial of naphthalene gives: K = 3, D = 16, D_2 = 19, D_3 = 4, and total number of canonical structures 42. In Figure 3, we give all resonance structure of naphthalene.

Wheland polynomials for benzene and several polycyclic benzenoids are given in Table 2.

It should be noted that the sum of all coefficients in the Wheland polynomials is the same for isomers. In other words, isomers have identical numbers of canonical structures.

The construction of the Wheland polynomials remains as the only procedure which allows partition of resonance structures into sets containing Kekulé structures, Dewar structures, etc.

F. The Permanental Polynomial

Permanental polynomial[50] is defined as

$$\text{per} \left| x\mathbf{I} - \mathbf{A} \right| = \sum_{k=0}^{N} c_k x^{N-k} \tag{61a}$$

where the symbols have the previous meaning. Note that $c_0 = 1$ by definition. Coefficients of the permanental polynomial can be calculated by means of the Sachs formula,[50,51]

$$c_k = (-1)^k \sum_{s \in S_k} 2^{c(s)} \tag{62}$$

where the summation is over all Sachs graphs s with k vertices. Other symbols have their previous meaning (see Chapter 5, Volume II, Section II).

Example

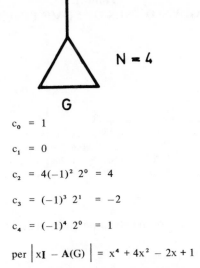

$$N = 4$$

G

$$c_0 = 1$$

$$c_1 = 0$$

$$c_2 = 4(-1)^2\ 2^0 = 4$$

$$c_3 = (-1)^3\ 2^1 = -2$$

$$c_4 = (-1)^4\ 2^0 = 1$$

$$\text{per}\left|x\mathbf{I} - \mathbf{A}(G)\right| = x^4 + 4x^2 - 2x + 1$$

The analysis of the coefficients of the permanental polynomial reveals some regularities, i.e.,

$$c_1 = 0$$

$$c_2 = \text{the number of edges in G}$$

$$-c_3 = \text{twice the number of } C_3 \text{ cycles in G}$$

$$c_N = K^2 \text{ (for bipartite graphs)}$$

For our purpose the last item is relevant.

Example

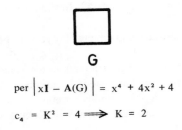

G

$$\text{per}\left|x\mathbf{I} - \mathbf{A}(G)\right| = x^4 + 4x^2 + 4$$

$$c_4 = K^2 = 4 \Longrightarrow K = 2$$

VI. ENUMERATION BASED ON THE LATTICE STRUCTURE OF FUSED BENZENOIDS

Gordon and Davis[52] considered catacondensed benzene rings and some finite lattice structures. By examining the underlying combinatorial nature of the problem, they arrived at the amazingly simple algorithm for enumerating the number of Kekulé structures which is illustrated below:

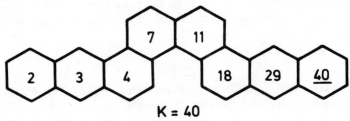

$$K = 40$$

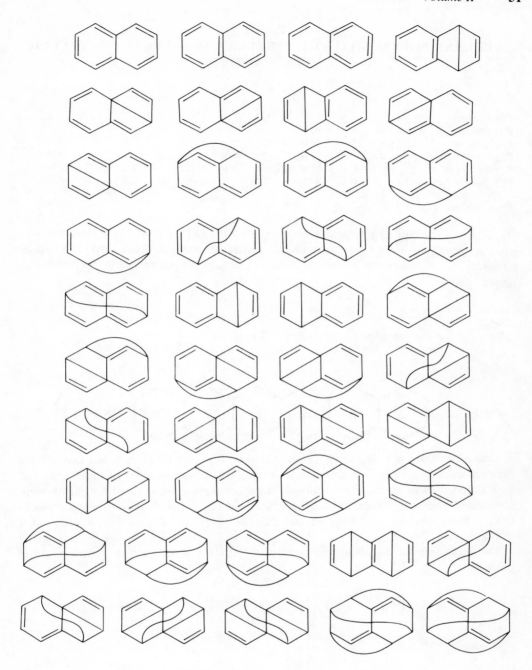

FIGURE 3.   Canonical resonance structures of naphthalene.

The procedure is to carry out a summation starting at either terminal hexagon, adding one figure to each hexagon. The numbers start with 2 (two Kekulé structures for a single benzene nucleus) and then 1 is added for any subsequent linearly fused hexagon. When a site with angularly fused ring, i.e., kink, appears, the quantity added to the previous figure is no longer unity, but the figure in the hexagon next to the last one preceding the kink. The procedure continues by adding this number for each subsequent linearly annelated ring until the next change of fusion direction comes when the process is repeated in the same fashion.

Table 2
WHELAND POLYNOMIALS OF BENZENE AND SEVERAL POLYCYCLIC
HYDROCARBONS

$P^w(G; x = 1)$

$P^w(benzene; x) = 2 + 3x$	5
$P^w(naphthalene; x) = 3 + 16x + 19x^2 + 4x^3$	42
$P^w(anthracene; x) = 4 + 48x + 150x^2 + 163x^3 + 58x^4 + 6x^5$	429
$P^w(phenanthrene; x) = 5 + 47x + 148x^2 + 165x^3 + 59x^4 + 5x^5$	429
$P^w(tetracene; x) = 5 + 110x + 649x^2 + 155x^3 + 164x^4 + 750x^5 + 138x^6 + 8x^7$	1979
$P^w(1,2\text{-benzanthracene}; x) = 7 + 112x + 642x^2 + 1551x^3 + 1654x^4 + 754x^5 + 135x^6 + 7x^7$	4862
$P^w(triphenylene; x) = 9 + 117x + 645x^2 + 1557x^3 + 1659x^4 + 744x^5 + 125x^6 + 6x^7$	4862

The scheme is not applicable to pericondensed systems or to a general structure. Nevertheless, when applicable, it is the fastest and simplest procedure available. Thus, it gives straight-forwardly, for example, the number of Kekulé structures for polyacenes by the following simple formula,

$$K = r + 1 \qquad (63)$$

where r is the number of hexagonal rings in a polyacene.

Example

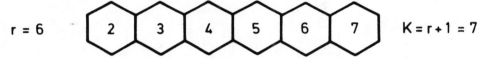

$r = 6$ | 2 | 3 | 4 | 5 | 6 | 7 | $K = r + 1 = 7$

Later, Yen[53] derived elegant equations, based on the work by Gordon and Davison for the enumeration of the Kekulé structures for several lattice systems of polynuclear benzenoid hydrocarbons. Four models for description of the lattice structure of fused benzenoids were considered. These are the *square* (parallelogram) *model* S[k,ℓ], the *symmetric circular model* C[k,ℓ], the *rectangular model* R[k,ℓ], and the *skew strip model* Z[k,ℓ]. They are illustrated in Figure 4.

The expression for calculating the number of Kekulé structures for the square model is given by,

$$K = \binom{k + \ell}{\ell} \qquad (64)$$

Example

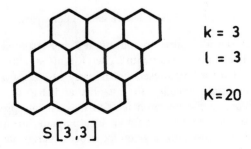

k = 3

l = 3

K = 20

s[3,3]

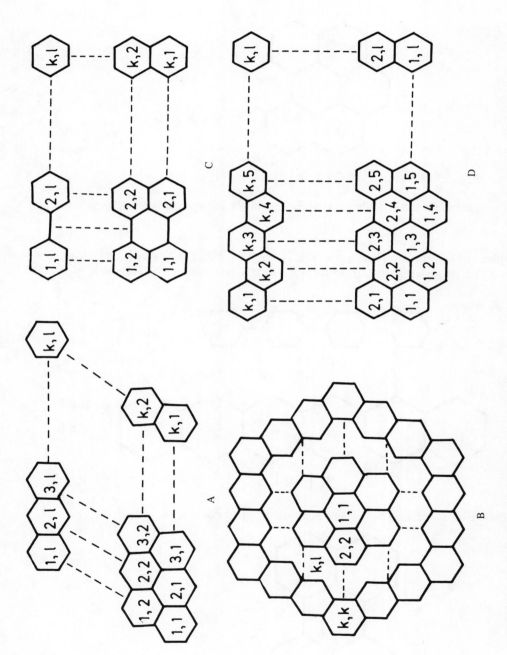

FIGURE 4. Lattice models of polynuclear benzenoid hydrocarbons. (A) The Square Model, S[k.l]; (B) the Circular Model, C[k.l]; (C) the Rectangular Model, R[k.l]; (D) the Skew Strip Model, Z[k.l].

The formula for hydrocarbons belonging to the circular model is as follows,

$$K = \frac{\binom{k+\ell}{\ell}\binom{k+\ell+1}{\ell}\ldots\ldots\binom{k+\ell+(k-1)}{\ell}}{\binom{\ell+1}{\ell}\ldots\ldots\binom{\ell+(k-1)}{\ell}} \tag{65}$$

Example

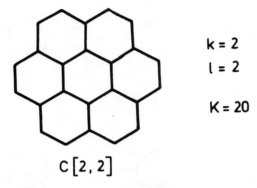

$$k = 2$$
$$l = 2$$

$$K = 20$$

C[2, 2]

The relation for the K of the rectangular model for polynuclear benzenoids is given by,

$$K = (\ell + 1)^k \tag{66}$$

which reduces for polyphenylenes, R[k,1], to $K = 2^k$, for polyacenes, R[1,ℓ], to $K = \ell + 1$, for polyrylenes, R[k,2], to $K = 3^k$, and for polyanthenes, R[k,3], to $K = 4^k$.

Example

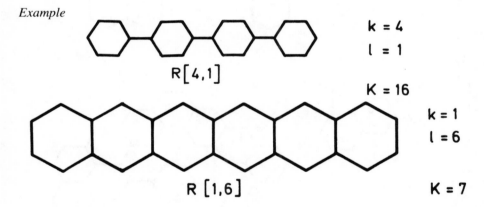

R[4,1]

$$k = 4$$
$$l = 1$$

$$K = 16$$

$$k = 1$$
$$l = 6$$

R [1,6]

$$K = 7$$

This result is identical to that one achieved by means of formula (63).

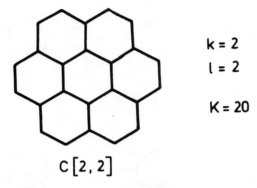

$$k = 2$$
$$l = 2$$

R [2,2] $$K = 9$$

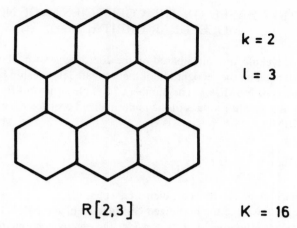

k = 2

l = 3

R[2,3]   K = 16

Equations (64), (65), and (66) can be collected in a general formula which embraces $S[k,\ell]$, $C[k,\ell]$, and $R[k,\ell]$ lattice models of polynuclear hydrocarbons,

$$K = \binom{\ell+b}{\ell}^c \prod_{q=1}^{k-\ell} \binom{k+\ell+q}{\ell} \Big/ \binom{\ell+a+q}{\ell} \tag{67}$$

where for (a) $S[k,\ell]$: $a = k$, $b = k$, $c = 1$, (b) $C[k,\ell]$: $a = 0$, $b = k$, $c = 1$, and (c) $R[k,\ell]$: $a = k$, $b = 1$, $c = k$.

The following are the formulae for the number of Kekulé structures of the skew strip model $Z[k,\ell]$. For $Z[k,1]$,

$$K = 2^{-(k+1)} \prod_{q=1}^{(k+2)/2} \binom{k+2}{2q-1} 5^{(q-1)}, \quad k = \text{even} \tag{68}$$

$$K = 2^{-(k+1)} \prod_{q=1}^{(k+3)/2} \binom{k+2}{2q-1} 5^{(q-1)}, \quad k = \text{odd} \tag{69}$$

For the $Z[1,\ell]$ series the expression for K is identical to that one for the polyacene series $R[1,\ell]$. The $Z[2,\ell]$ series has the number of Kekulé structures given by,

$$K = (\ell+1)(\ell+2)/2 \tag{70}$$

The next sequence $Z[3,\ell]$ leads to,

$$K = (\ell+1)(\ell+2)(2\ell+3)/6 \tag{71}$$

Example

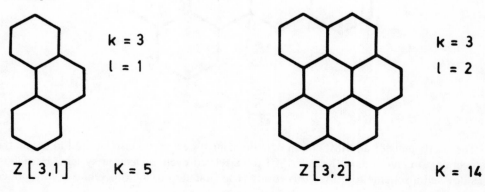

k = 3

l = 1

Z[3,1]   K = 5

k = 3

l = 2

Z[3,2]   K = 14

VII. ENUMERATION BASED ON THE COEFFICIENTS OF NONBONDING MOLECULAR ORBITALS

The enumeration of Kekulé structures based on the coefficients of nonbonding molecular orbitals NBMOs is founded on the recognition of the intimate relationship between the value of K and the coefficients of NBMOs.[54] The number of Kekulé structures for an *odd* alternant is equal to the sum of the absolute values of the unnormalized coefficients of the nonbonding molecular orbital, NBMO.

$$\sum_{i=1}^{N} \left| c_{0i} \right| = K \tag{72}$$

where the summation is over all the coefficients c_{0i} ($i = 1, 2, \ldots, N$) of the NBMO. It is not difficult to write down the unnormalized coefficients of the NBMO by means of the *zero-sum rule* of Longuet-Higgins.[55] Let us consider the eigenvalue equation,

$$\mathbf{C_i A} = x_i \mathbf{C_i} \tag{73}$$

In the case of $N_0 \neq 0$, $x_i = 0$ and $\mathbf{C_i} = \mathbf{C_0} = $ NBMO,

$$\mathbf{C_0 A} = 0 \tag{74}$$

or in scalar form,

$$\sum_{j \to k} c_{0j} A_{jk} = 0; k = 1, 2, \ldots, N \tag{75}$$

where the summation is over all vertices *j* joined to the vertex *k*.

One set of atoms (unstarred atoms) have zero coefficients in the NBMO, the other set (starred atoms) have simple integers chosen in such a way that the sum of all nonvanishing coefficients around an unstarred atom is zero.[56]

Example

$$K = \sum_{i=1}^{23} c_{0i} = 48$$

The coefficients of a NBMO of an odd alternant system could also be used to obtain the (algebraic) structure count (K and ASC) for a related even π-electron system. Herndon[35] has devised a scheme to utilize this result. The first step in the procedure is to produce an odd alternant from the even alternant by deleting one carbon atom and adjacent bonds from the even system. The second step is to write a NBMO by one of the available methods.[54-57] Then, the sum of absolute values of the unnormalized coefficients c_{0r} at points r adjacent to the deleted atom s is equal to the K of the parent compound,

$$\sum_{r \to s} \left| c_{0r} \right| = K \tag{76}$$

In addition, the sum of the coefficients defines the algebraic structure count,

$$\sum_{r \to s} c_{0r} = ASC \tag{77}$$

In the case of benzenoids these two relations, of course, are identical.

Example

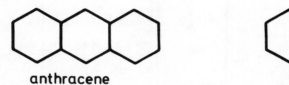

anthracene   **biphenylene**

(a) Creation of odd-alternant structure by excising an arbitrary atom s from the molecule.

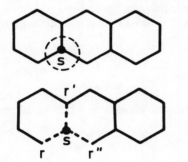
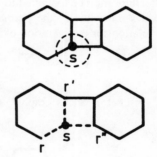

(b) Construction of the unnormalized NBMO

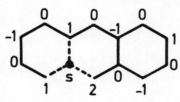
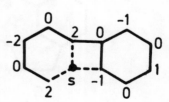

(c) Application of Equation (76) gives

$$K \text{ (anthracene)} = c_{0r} + c_{0r'} + c_{0r''} = 4$$

$$K \text{ (biphenylene)} = \left| c_{0r} \right| + \left| c_{0r'} \right| + \left| c_{0r''} \right| = 5$$

(d) Application of Equation (77) gives

$$ASC \text{ (anthracene)} = c_{0r} + c_{0r'} + c_{0r''} = 4$$

$$ASC \text{ (biphenylene)} = c_{0r} + c_{0r'} + c_{0r''} = 3$$

VIII. ENUMERATION BASED ON THE PERMANENT OF THE ADJACENCY MATRIX

The permanent of the adjacency matrix, per $\mathbf{A}(G)$, is related to the number of Kekulé valence structures. It has been shown that the square of the number of Kekulé structures of alternant hydrocarbons is equal to per $\mathbf{A}(G)$

$$K^2 = \text{per } \mathbf{A}(G) \qquad\qquad (78)$$

Example

$$\mathbf{A}(G) = \begin{bmatrix} 0 & 1 & 0 & 0 & 0 & 1 \\ 1 & 0 & 1 & 0 & 0 & 0 \\ 0 & 1 & 0 & 1 & 0 & 0 \\ 0 & 0 & 1 & 0 & 1 & 0 \\ 0 & 0 & 0 & 1 & 0 & 1 \\ 1 & 0 & 0 & 0 & 1 & 0 \end{bmatrix}$$

per $\mathbf{A}(G) = 4 \implies K = 2$

This procedure may be extended to embrace nonalternant hydrocarbons. In that case one needs, besides the value of the permanent of $\mathbf{A}(G)$, to make the use of Sachs formula.[51] Then, the relation for enumeration of K is given by,[17]

$$K^2 = \text{per } \mathbf{A}(G) - \sum_{s \in S_N^{\text{odd}}} 2^{r(s)} \qquad\qquad (79)$$

where S_N^{odd} is a set of Sachs graphs[23] s with N vertices containing r odd-membered rings.

Example

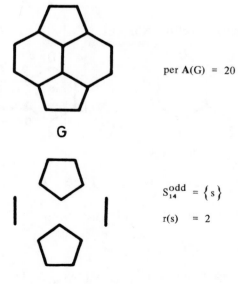

per $\mathbf{A}(G) = 20$

$S_{14}^{\text{odd}} = \{s\}$

$r(s) = 2$

$K^2 = 20 - 2^2 \implies K = 4$

IX. ENUMERATION BASED ON THE GRAPH DECOMPOSITION

All approaches for the enumeration of Kekulé structures so far mentioned analyzed a given conjugated system and produced a needed result, but said little about the relationship between the K value of a system and the corresponding K values of its constituent parts. Here we will discuss a scheme which shows how one may obtain the number of Kekulé structures of a composite system from the corresponding numbers of its constituents.[38] In the catacondensed systems, the following structural parts are permitted for the immediate environment at the site of fusion of two fragments A and B,

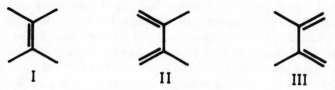

The fusion site is either a double bond (I), or a single bond with both adjacent double bonds in a *cis* formation, (II) and (III). Let *s* (and *d*) be the number(s) of Kekulé structures of a fragment in which the fusion site is a single (and double) bond. The number of Kekulé structures of a system consisting of two fragments A and B is then given by,

$$K = d_A d_B + d_A s_B + s_A d_B \qquad (80)$$

The value of K is, of course, independent of the way the system is fragmented into A and B.

Example

Two naphthalene units can combine in four structures

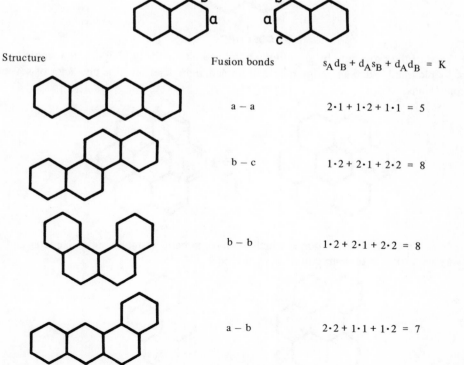

Structure	Fusion bonds	$s_A d_B + d_A s_B + d_A d_B = K$
	a — a	$2 \cdot 1 + 1 \cdot 2 + 1 \cdot 1 = 5$
	b — c	$1 \cdot 2 + 2 \cdot 1 + 2 \cdot 2 = 8$
	b — b	$1 \cdot 2 + 2 \cdot 1 + 2 \cdot 2 = 8$
	a — b	$2 \cdot 2 + 1 \cdot 1 + 1 \cdot 2 = 7$

A more general counting procedure for Kekulé valence structures is based on the relationship between the number of Kekulé structures of the parent conjugated molecule and of its constituting fragments.[38] The procedure consists of the reduction of a given conjugated hydrocarbon (graph) to smaller fragments (subgraphs) of known number of Kekulé structures. The steps in the procedure are as follows: (1) a bond in a given molecule is selected. In some Kekulé structures it will appear as a double bond and in others as a single bond. Thus, two structures are initially generated: one with the double bond assigned and the other with a single bond assigned; (2) in these two structures, the immediate environment is considered. In the structure with a double bond, the bonds adjacent to the double bond must be single; they cannot affect the distribution of double bonds in the remaining part of the molecule and hence can be eliminated together with the double bond from the further consideration. In the structure with a single bond, the single bond must be connected by the double bonds to the rest of the molecule. These double bonds must have single bonds as nearest neighbors. Thus, one can assign immediately the bond type to at least five bonds, all of which can be deleted from the molecule since the attachment through a single bond does not change the Kekulé structure count of the leftover part. In some cases one may assign the bond type even to more bonds than just five of them (see example below); (3) in this way, the initial molecule is decomposed into smaller parts and the process continues until one arrives at parts with known numbers of Kekulé structures. Therefore,

$$K \text{ (parent conjugated molecule)} = \sum_i K_i \text{ (constituting fragments)} \tag{81}$$

Example

coronene

(i) We start the procedure by selecting a bond and alternatively assign to it a single bond and a double bond. Such an assignment will immediately determine the character of several adjacent bonds.

 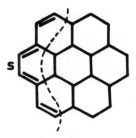

(ii) Now we erase all those bonds which we have already assigned bond character and in this way we arrive at two structures of smaller size.

(iii) The process continues considering both structures.

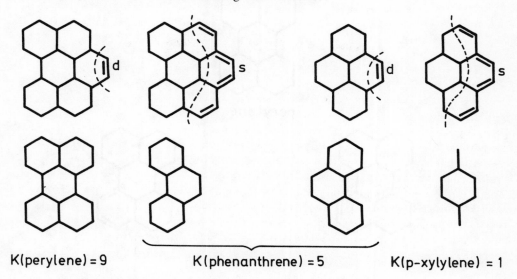

K(perylene) = 9   K(phenanthrene) = 5   K(p-xylylene) = 1

The decomposition of perylene is shown in Figure 5.

(iv) K(coronene) = K(perylene) + 2K(phenanthrene) + K(p-xylylene) = 20

This method is very usable for large conjugated systems, because it is rapid and does not require any calculations. The decomposition of a graph can be accelerated by eliminating all edges in a graph for which the unique assignment of the bond type (i.e., single or double) is possible. This can be done by following a set of rules.[38,58] These graph decomposition rules are

Rule 1 — Acyclic groups with an even number of bonds can be deleted because they cannot alter the distribution of single and double bonds in the cyclic part of a molecule.

Example

K = 3   K = 3

Rule 2 — Exocyclic double bond(s) can be deleted together with the adjacent ring bonds since they must be single.

Example

K = 2   K = 2

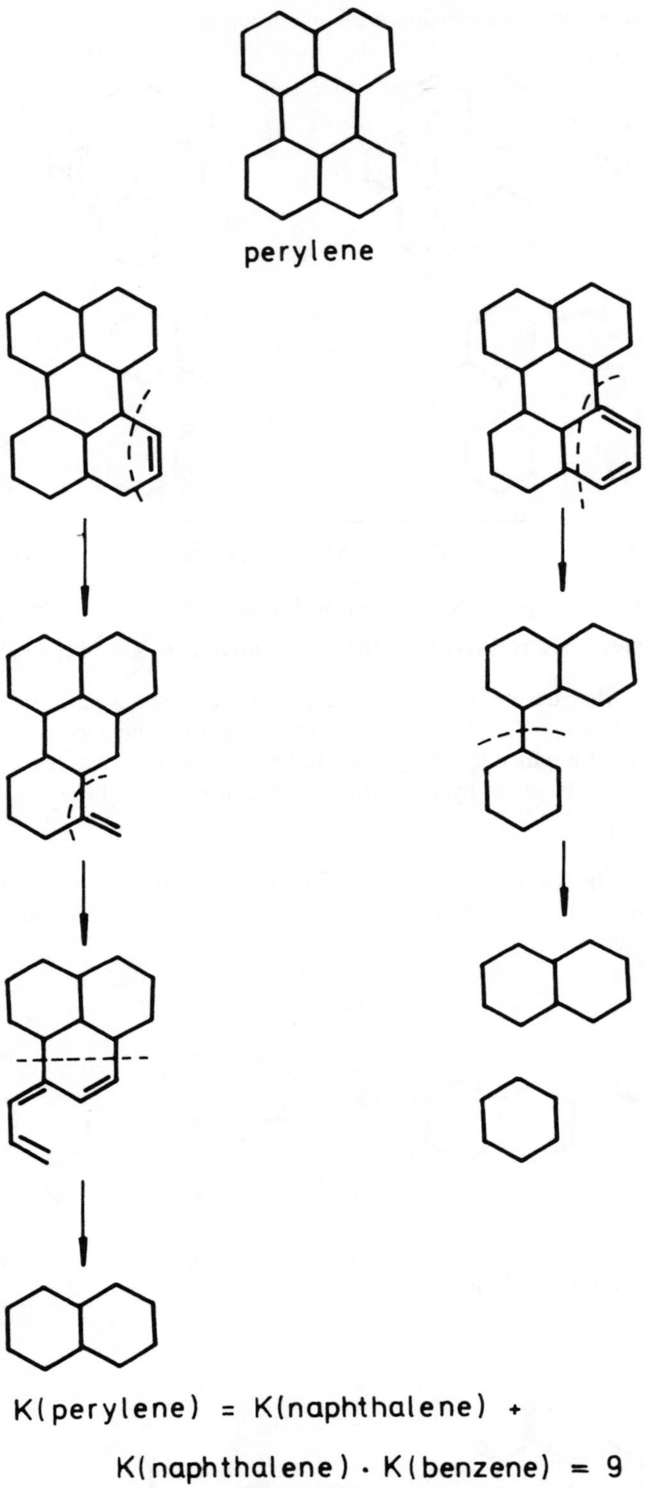

perylene

$$K(\text{perylene}) = K(\text{naphthalene}) +$$
$$K(\text{naphthalene}) \cdot K(\text{benzene}) = 9$$

FIGURE 5.   The decomposition of perylene in the constituent fragments.

Rule 3 — A parent compound consisting of units joined by a single bond may be partitioned into these units by deleting the connecting bond. The number of Kekulé structures of a parent hydrocarbon is then given as a product of the corresponding numbers of constituting units.

Example

K(parent hydrocarbon) = K(anthracene)·K(naphthalene) = 12

Rule 4 — A bridge of *n* double bonds can be removed from the parent structure.

Example

X. STRUCTURE COUNT FOR CONJUGATED RADICAL CATIONS

The use of the Herndon's Equation (36) for the calculation of the first ionization potential[32] of conjugated molecules requires the values of the algebraic structure count for the neutral molecule and for the corresponding radical cation. The ways to obtain the algebraic structure count for the conjugated hydrocarbons are presented in preceding pages. The Randić's relation,[34]

$$ASC(G^{\cdot +}) = 2SC(G)\, n_{c=c}(G) + 2\, D(G) \tag{82}$$

may serve for obtaining the algebraic structure count for conjugated radical cations. $D(G)$ stands for the number of Dewar structures. Note that $SC(G) = K(G)$. Since for neutral closed-shell hydrocarbons,

$$n_{C=C}(G) = N(G)/2 \tag{83}$$

Equation (82) may be presented in different form,

$$ASC(G^{\cdot +}) = SC(G)\cdot N(G) + 2\, D(G) \tag{84}$$

Relations (82) or (84) are not easy to use, because the number of Dewar structures is not a readily available datum. However, the Dewar structures count is somewhat simpler for benzenoid structures than for those containing $4m$ rings.

One way to obtain $D(G)$ values of benzenoids is by means of the Wheland polynomials. The number of Dewar structures is equal to the *second* coefficient of the Wheland polynomial (see Section V.E),

$$D(G) = W(G; k=1) \tag{85}$$

Example

$$n_{C=C} = 7; \quad N = 14$$

$$SC(G) = 5$$

$$D(G) = 47 \text{ (see Table 2)}$$

G = phenanthrene

$$ASC(G^{.+}) = 14 \cdot 5 + 2 \cdot 47 = 164$$

There is another avenue open to enumerate valence structures of benzenoid radical cations.[59] This method makes use of the $a_{N-2}^{ac}(G)$ coefficient of the acyclic polynomial of a benzenoid molecule. (Acyclic polynomial is defined in Chapter 1, Section I). It has been found, by considering the construction of the $a_{N-2}^{ac}(G)$ coefficient, that the absolute value of this coefficient is given by

$$\left| a_{N-2}^{ac}(G) \right| = SC(G) \cdot n_{C=C}(G) + D(G) \tag{86}$$

Formula (86) is obtained by examining in detail the building of the $a_{N-2}^{ac}(G)$ coefficient by means of the Sachs formula. Only graphs with N = even can be considered. The K_2 subgraphs cover a given benzenoid graph G in such a way that N − 2 vertices are covered, while two vertices are always left free. However, these uncovered vertices are either adjacent or separated by a certain number of K_2 fragments. This is illustrated below for the case of naphthalene.

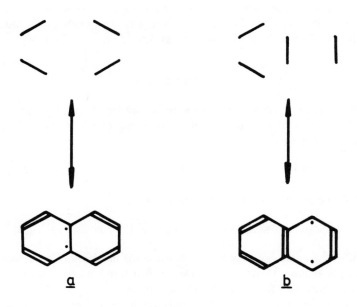

a   **b**

Inspection of these structures reveals that the structure *a* needs only *one* additional K_2 fragment to generate a Kekulé form, while the structure *b* resembles a Dewar structure missing the "long" bond connecting two nonadjacent vertices, and, of course, cannot be used for generating Kekulé forms. It turns out that the number of structures a is simply equal to the number of permutations of $1/2(N-2)K_2$ fragments on the graphs with N vertices, always leaving two *adjacent* vertices unconnected. Since each permutation leads to a structure which upon addition of a K_2 graph produces a single Kekulé form, the total number of structures a is given by $SC(G) \cdot n_{C=C}(G)$.

Example

$$N = 10; \quad n_{C=C}(G) = 5$$
$$SC(G) = 3$$

(Label below each structure a denotes to which Kekulé form it leads)

$$SC(G) \cdot n_{C=C}(G) = 3 \cdot 5 = 15$$

The number of structures b is equal to the number of permutations $1/2(N-2)K_2$ fragments on the graphs with N vertices, always leaving two *nonadjacent* vertices, an even number of centers apart, disconnected. Since each permutation leads to a structure which upon addition of a "long" bond generates a single Dewar form of a molecule, the total number of structures b is equal to $D(G)$.

Example

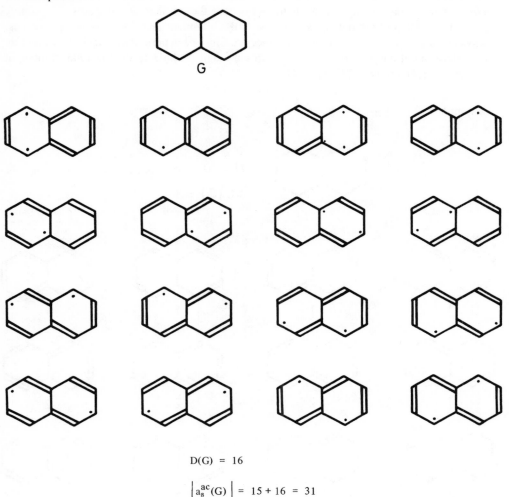

$$D(G) = 16$$

$$\left| a_8^{ac}(G) \right| = 15 + 16 = 31$$

The rearrangement of formula (86) into a somewhat different form leads to the expression for the enumeration of Dewar structures of a given conjugated molecule G,

$$D(G) = \left| a_{N-2}^{ac}(G) \right| - SC(G) \cdot n_{C=C}(G) \tag{87}$$

Finally, by combining Equations (82) and (87) we arrive at a rather simple expression for the enumeration of valence structures for radical cations,

$$ASC(G^{\cdot +}) = 2 \left| a_{N-2}^{ac}(G) \right| \tag{88}$$

The validity of Equation (88) may be confirmed by considering the generation of radical cations from structures *a* and *b*.

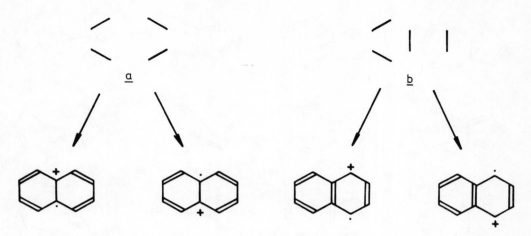

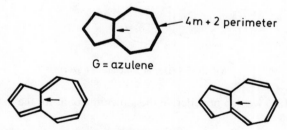

Every structure *a* and every structure *b* always generates *two* different forms of radical cations.

Let us now obtain the structure count for the naphthalene radical cation, since above we have derived the value of the coefficient $a_{N-2}^{ac}(G)$,

$$SC(G^{\cdot+}) = 2|a_8^{ac}(G)| = 2\cdot31 = 62$$

Equations (86), (87), and (88), besides being used for benzenoids, may be also used directly for azulenoids. This is so because azulenoids have the $4m + 2$ periphery germane to that of benzenoids, free of disruption from the transannular bonds which are all essentially single bonds. (Transannular bonds appear as single bonds in all Kekulé structures of the compound).

Example

G = azulene — 4m + 2 perimeter

(Essential single bond is indicated by arrows.)

The acyclic polynomial of azulene is given by,

$$P^{ac}(G; x) = x^{10} - 11x^8 + 41x^6 - 61x^4 + 31x^2 - 2$$

Hence,

$$SC(G) = 2$$

i.e., there are two Kekulé valence structures in azulene. Using the value of the coefficient $a_{N-2}^{ac}(G) = 31$ and Equation (87) we obtain,

$$D(G) = 21$$

i.e., there are 21 singly excited Dewar-type valence structures in azulene which are depicted in Figure 6.

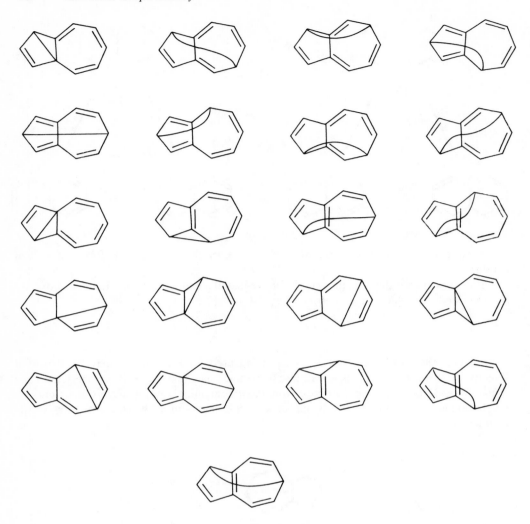

FIGURE 6.   Dewar structures of azulene.

From Equation (88) we can now derive the structure count for the azulene radical cation,

$$SC(G^{\cdot +}) = 2 \cdot 31 = 62$$

In order to treat cyclobutadienoids by this procedure, the Randić's, relation (82) should be modified accordingly,

$$ASC(G^{\cdot +}) = 2[SC(G) \cdot n_{C=C}(G) - L + CDSC(G)] \qquad (89)$$

where L is the number of structures *a* with opposite parity, while CDSC(G) stands for a corrected Dewar structure count. The parity correction is taken into account only when structures *a* contain 4*m* ring components containing exactly 4*m* paired π-electrons. Only some of the structures *a* will contain fully conjugated 4*m* rings. There will be exactly 4*m*/2 of these structures of positive parity and exactly 4*m*/2 of these structures of negative parity. Therefore, the structures with the opposite parity will not contribute to Equation (89) and they should be taken away from it. Thus, the first part of relation (89) equals the alegbraic structure count of structures *a*,

$$ASC(a) = SC(G) \cdot n_{C=C}(G) - L \tag{90}$$

The L correction can be simply obtained by considering those Kekulé structures of a cyclobutadienoid which contain four paired π-electrons in the 4-membered cycle. Then the total number of double bonds in the adjacent rings, not participating in the π system of 4-membered rings, in each such a Kekulé structure is L. For example biphenylene has 5 Kekulé structures. Among these only two contain fully conjugated 4-membered ring.

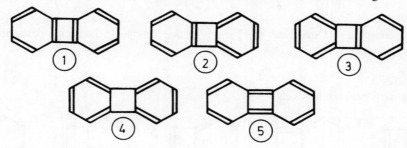

Kekulé structures ① and ⑤ contain fully conjugated 4-membered cycle. The total number of double bonds in the adjacent 6-membered rings not participating in the conjugation in the 4-cycle in both structures is L = 4 + 4 = 8. Thus, the ASC(a) of biphenylene is 22.

Since the Dewar structures of cyclobutadienoids also have parity, CDSC(G) values should be considered instead of D(G) for this class of conjugated radical cations. CDSC(G) is defined as,

$$CDSC(G) = D^+(G) - D^-(G) \tag{91}$$

where $D^+(G)$ and $D^-(G)$ are Dewar structures of positive and negative parity, respectively.

Example
Enumeration of valence structures for benzocyclobutadiene radical cation

$$N(G) = 8$$
$$n_{C=C}(G) = 4$$
$$SC(G) = 3$$

G

Structures a
(Parity sign is given brackets)

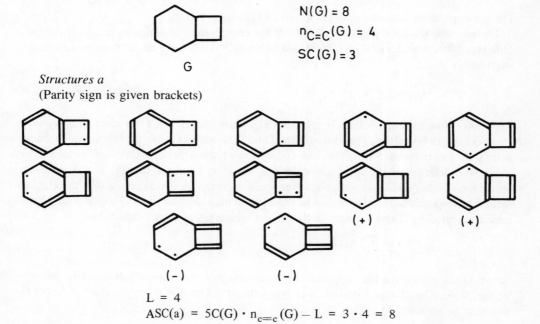

(−)   (−)   (+)

L = 4
$$ASC(a) = 5C(G) \cdot n_{c=c}(G) - L = 3 \cdot 4 = 8$$

This result may also be achieved by considering only Kekulé structures of benzocyclobutadiene which contain fully conjugated 4-membered cycle. There are three Kekulé structures of benzocyclobutadiene.

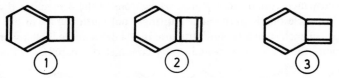

By considering Kekulé structures ② and ③ and following the recipe described above we obtain L = 2 + 2 = 4.

Structures b
(Parity sign is given brackets)

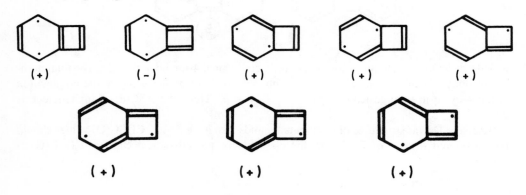

$$CDSC(G) = D^+(G) - D^-(G) = 7 - 1 = 6$$

The valence structure count for the benzocyclobutadiene radical cation is,

$$ASC(G^+) = 2[ASC(a) + CDSC(G)] = 2\,(8 + 6) = 28$$

The corresponding structures are presented in Figure 7.

There is also the third procedure available for enumeration of conjugated radical cations. This procedure was devised by Eilfeld and Schmidt.[60] They have based their procedure on the relation,[13]

$$[ASC(G)]^2 = \prod_{i=1}^{N} |x_i| \qquad (92)$$

They have also used the Randić's relation (82). Their target was to produce an easy way for obtaining D(G) values. The idea of Eilfeld and Schmidt was to use the Hückel (adjacency) matrix in the reduced form. The reduced forms of the Hückel matrix are obtained by successive deletion of the rows and columns, u and v, from the matrix. If one wants to obtain the number of Dewar structures then the deletion of rows and columns in the Hückel matrix corresponds to the removal of nonadjacent centers u and v from G which would have been connected by "long" bonds in the Dewar structures of a molecule. Thus,

$$D(G) = \sum_{u<v} ASC(G\text{-}u\text{-}v) \qquad (93)$$

where G-u-v stands for the subgraph corresponding to a structure which upon the addition of the "long" bond would become a Dewar structure. The ASC(G-u-v) values of larger G-u-v may be obtained by,

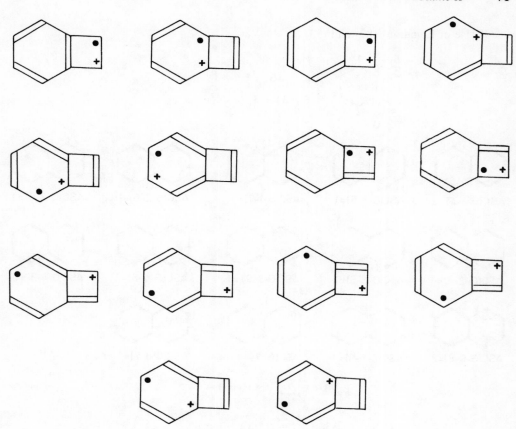

FIGURE 7. Half of the valence structures for benzocyclobutadiene radical cation. The other half of structures can be simply obtained by exchanging the positions of · and +.

$$[ASC(G\text{-}u\text{-}v)]^2 = \prod_{i=1}^{L} |x_i| \qquad (94)$$

where $L = N - 2$.

Since we are interested in the total number of cation radicals, all pairs of u, v vertices should be successively deleted from G regardless whether or not they are adjacent. Then,

$$ASC(G^{\cdot+}) = 2 \sum_{u<v} ASC(G\text{-}u\text{-}v)' \qquad (95)$$

where (G-u-v)′ stands for the subgraphs obtained by successive elimination of the (u, v) pairs of vertices from G. The structure count for each subgraph is then found, if needed, by means of (92) For benzenoids, Equation (95) gives the same result as Equation (93) in combination with Equation (82).

Example

Structure count for naphthalene radical cation using the Eilfeld-Schmidt procedure

(a) Use of Equations (82) and (93)

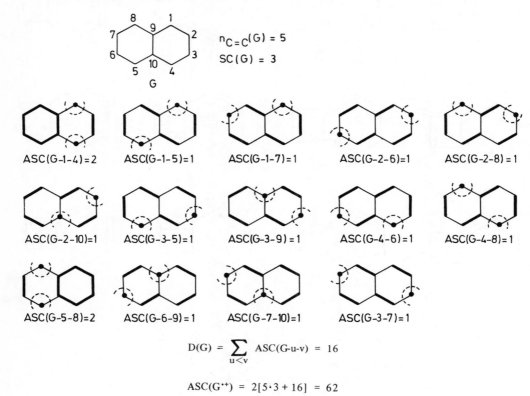

$n_{C=C}(G) = 5$

$SC(G) = 3$

$$D(G) = \sum_{u<v} ASC(G\text{-}u\text{-}v) = 16$$

$$ASC(G^{\cdot+}) = 2[5 \cdot 3 + 16] = 62$$

(b) Use of Equation (95)

(G-u-v)′ subgraphs and ASC(G-u-v)′ values (numbers below structures)

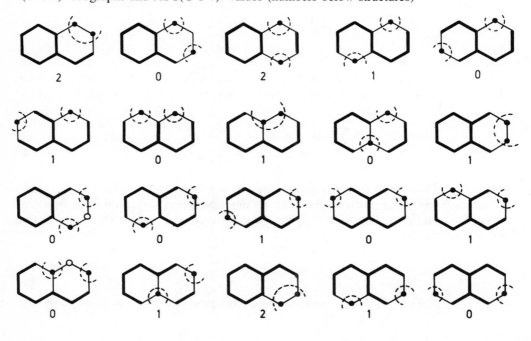

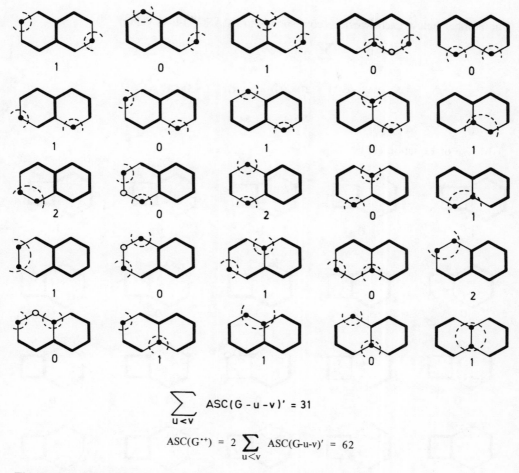

$$\sum_{u<v} ASC(G-u-v)' = 31$$

$$ASC(G^{\cdot+}) = 2 \sum_{u<v} ASC(G-u-v)' = 62$$

The Eilfeld-Schmidt procedure is also applicable to cyclobutadienoids.

Example

Structure count for benzocyclobutadiene radical cation using the Eilfeld-Schmidt procedure

(a) Use of Equations (82) and (93)

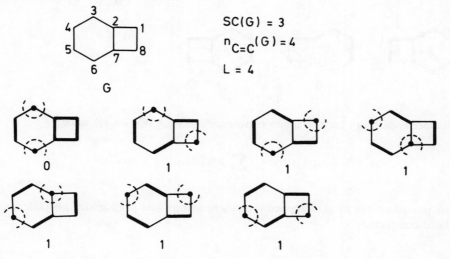

(The number below each structure corresponds to its ASC(G-u-v) value)

$$D(G) = \sum_{u<v} ASC(G\text{-}u\text{-}v) = 6$$

$$ASC(G^{\cdot+}) = 2[SC(G) \cdot n_{C=C}(G) - L + D(G)] =$$

$$2[3 \cdot 4 - 4 + 6] = 28$$

(b) Use of Equation (95)

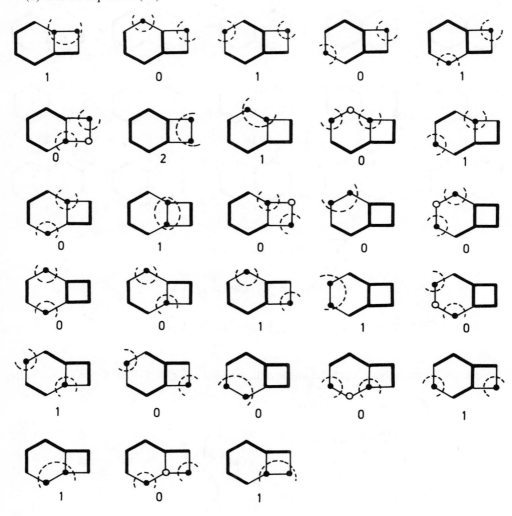

(Number underneath each structure corresponds to its ASC(G-u-v)′ value)

$$ASC(G^{\cdot+}) = 2\sum_{u<v} ASC(G\text{-}u\text{-}v)' = 2 \cdot 14 = 28$$

This procedure can be also applied directly to azulenoids, pentalenoids, heptalenoids, and related compounds.

Example

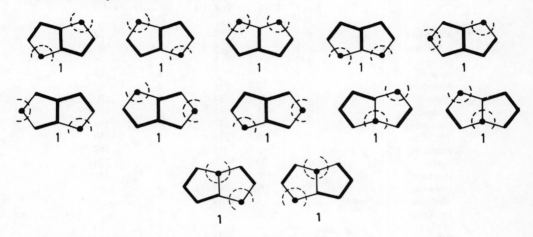

$$n_{C=C}(G) = 4$$
$$SC(G) = 2$$

(a) Use of Equations (82) and (93)

(Number below each structure corresponds to its ASC(G-u-v) value)

$$D(G) = \sum_{u<v} ASC(G\text{-}u\text{-}v) = 12$$

$$ASC(G^{\cdot+}) = 2[SC(G)\cdot n_{C=C}(G) + D(G)] = 2[2\cdot4 + 12] = 40$$

(b) Use of Equation (95)

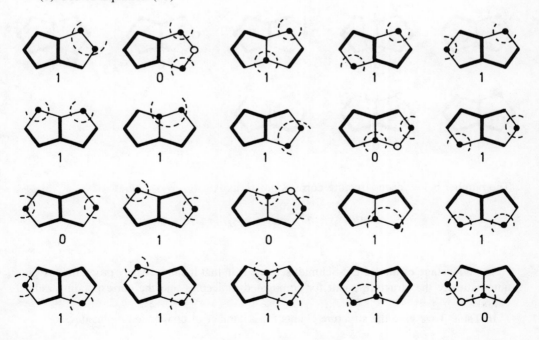

Table 3

THE NUMBER OF KEKULÉ STRUCTURES, K(G), THE ALGEBRAIC STRUCTURE COUNT, ASC(G), THE NUMBER OF DEWAR STRUCTURES, D(G), THE CORRECTED DEWAR STRUCTURE COUNT, CDSC(G), FOR NEUTRAL HYDROCARBONS AND THE ALGEBRAIC STRUCTURE COUNT, ASC(G·⁺), FOR THE RADICAL CATIONS

G	K(G)	D(G)	CDSC(G)	ASC(G)	ASC(G·⁺)
o-Xylylene	1	7	7	1	22
p-Xylylene	1	6	6	1	20
Styrene	2	6	6	2	28
Benzene	2	3	3	2	18
Naphthalene	3	16	16	2	62
Biphenyl	4	21	21	4	90
Anthracene	4	48	48	4	152
Phenanthrene	5	47	47	5	164
Tetracene	5	110	110	5	310
Pentacene	6	215	215	6	562
Pyrene	6	79	79	6	254
Perylene	9	178	178	9	536
Coronene	20	504	504	20	1488
Methylene-cyclopropene	1	2	2	1	8
Fulvene	1	5	5	1	16
Heptafulvene	1	9	9	1	26
Azulene	2	21	21	2	62
Pentalene	2	12	12	0	40
Heptalene	2	33	33	0	90
Cyclobutadiene	2	0	0	0	8
Acenaphthylene	3	38	38	3	112
Benzocyclobutadiene	3	8	6	1	28
Biphenylene	5	26	22	3	88

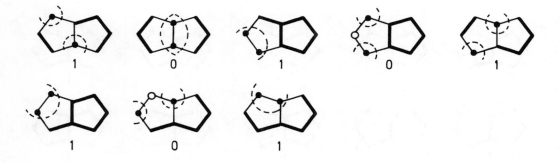

(The number below each structure corresponds to its ASC(G-u-v)′ value)

$$ASC(G^{·+}) = 2\sum_{u<v} ASC(G\text{-}u\text{-}v)' = 2·20 = 40$$

The advantage of the Eilfeld-Schmidt scheme is in that it can be easily programed to give automatically the structure count for conjugated molecules and the corresponding radical cations.

In Table 3 we give the structure counts for a number of conjugated molecules.

REFERENCES

1. **Morrison, R. T. and Boyd, R. N.,** *Organic Chemistry,* Allyn & Bacon, Boston, 1973, 319.
2. **Ternay, A. L.,** *Contemporary Organic Chemistry,* W. B. Saunders, Philadelphia, 1979, 568.
3. **Kekulé, A.,** *Bull. Soc. Chim., Fr.,* 3, 98, 1865; *Justus Liebigs Ann. Chem.,* 137, 129, 1866.
4. **Pauling, L.,** *The Nature of Chemical Bond,* 2nd ed., Cornell University Press, Ithaca, N.Y., 1948.
5. **Wheland, G. W.,** *The Theory of Resonance and Its Application to Organic Chemistry,* John Wiley & Sons, New York, 1953, sixth printing.
6. **Fries, K.,** *Justus Liebigs Ann. Chem.,* 454, 121, 1927; see also **Kekulé, A.,** *Justus Liebigs Ann. Chem.,* 162, 77, 1872.
7. **Pauling, L.,** *Acta Cryst.,* B 36, 1898, 1980; **Herndon, W. C.,** *Isr. J. Chem.,* 20, 270, 1980.
8. **Dewar, M. J. S.,** *Chem. Br.,* 11, 97, 1975.
9. **Gutman, I., Trinajstić, N., and Wilcox, C. F., Jr.,** *Tetrahedron,* 31, 143, 1975; 31, 147, 1975.
10. **Clar, E., Kemp, W., and Stewart, D. G.,** *Tetrahedron,* 3, 325, 1958.
11. **Herndon, W. C.,** *J. Am. Chem. Soc.,* 95, 2404, 1973; *Thermochim. Acta,* 8, 225, 1974; **Herndon, W. C. and Párkányi, C.,** *J. Chem. Educ.,* 53, 689, 1976.
12. **Randić, M.,** *Chem. Phys. Lett.,* 38, 68, 1976; *J. Am. Chem. Soc.,* 99, 444, 1977; *Tetrahedron,* 33, 1905, 1977; *Int. J. Quantum Chem.,* 17, 549, 1980.
13. **Dewar, M. J. S. and Longuet-Higgins, H. C.,** *Proc. Soc. London, Ser. A,* 214, 482, 1952; **Ham, N. S. and Ruedenberg, K.,** *J. Chem. Phys.,* 29, 1215, 1958; **Cvetković, D., Gutman, I., and Trinajstić, N.,** *J. Chem. Phys.,* 61, 2700, 1974.
14. **Wilcox, C. F., Jr.,** *Tetrahedron Lett.,* 795, 1968.
15. **Gutman, I. and Trinajstić, N.,** *Croat. Chem. Acta,* 47, 35, 1975.
16. **Gutman, I., Randić, M., and Trinajstić, N.,** *Rev. Roum. Chim.,* 23, 383, 1978.
17. **Cvetković, D., Gutman, I., and Trinajstić, N.,** *Chem. Phys. Lett.,* 16, 614, 1972.
18. **Jacobson, I.,** *Chem. Scripta,* 4, 30, 1974.
19. **Wilcox, C. F., Jr.,** *J. Am. Chem. Soc.,* 91, 2732, 1969.
20. **Coulson, C. A.,** *Proc. Cambridge Phil. Soc.,* 46, 202, 1950.
21. **Gutman, I. and Trinajstić, N.,** *Chem. Phys. Lett.,* 20, 257, 1973.
22. **Günthard, H. H. and Primas, H.,** *Helv. Chim. Acta,* 39, 1645, 1956.
23. **Graovac, A., Gutman, I., Trinajstić, N., and Živković, T.,** *Theor. Chim. Acta,* 26, 67, 1972; see also **Gutman, I. and Trinajstić, N.,** *Croat. Chem. Acta,* 45, 539, 1973.
24. **Trinajstić, N.,** in *Semiempirical Methods of Electronic Structure Calculation. Part A: Techniques,* Vol. 7, Segal, G. A., Ed., Plenum Press, New York, 1977, 1.
25. **Schaad, L. J. and Hess, B. A., Jr.,** *J. Am. Chem. Soc.,* 94, 3068, 1972.
26. **Gutman, I., Milun, M., and Trinajstić, N.,** *J. Am. Chem. Soc.,* 94, 1692, 1977.
27. **Gutman, I. and Herndon, W. C.,** *Chem. Phys. Lett.,* 34, 387, 1975.
28. **Herndon, W. C. and Ellzey, M. L., Jr.,** *J. Am. Chem. Soc.,* 96, 6631, 1974.
29. **Herndon, W. C.,** *J. Am. Chem. Soc.,* 96, 7605, 1974.
30. **Pauling, L.,** *J. Chem. Phys.,* 1, 280, 1933.
31. **Dewar, M. J. S. and de Llano, C.,** *J. Am. Chem. Soc.,* 91, 789, 1969.
32. **Herndon, W. C.,** *J. Am. Chem. Soc.,* 98, 887, 1976.
33. **Swinborne-Sheldrake, R., Herndon, W. C., and Gutman, I.,** *Tetrahedron Lett.,* 755, 1975.
34. **Randić, M.,** private communication.
35. **Herndon, W. C.,** *Tetrahedron,* 29, 3, 1973.
36. **Herndon, W. C.,** *J. Chem. Educ.,* 51, 10, 1974.
37. **Cvetković, D., Gutman, I., and Trinajstić, N.,** *Theor. Chim. Acta,* 34, 129, 1974.
38. **Randić, M.,** *J. Chem. Soc. Faraday Trans. 2,* 232, 1976; **Gutman, I. and Trinajstić, N.,** *Croat. Chem Acta,* 45, 423, 1973.
39. **Polansky, O. E. and Gutman, I.,** *Math. Chem. (Mülheim/Ruhr),* 8, 269, 1980.
40. **Džonova-Jerman-Blažić, B. and Trinajstić, N.,** Computers & Chemistry, 6, 121, 1982; *Croat. Chem. Acta,* 55, 347, 1982.
41. **Hosoya, H.,** *Bull. Chem. Soc. Jpn.,* 44, 2332, 1971.
42. **Hosoya, H. and Yamaguchi, T.,** *Tetrahedron Lett.,* 4659, 1975; see also **Ohkami, N., Motoyama, A., Yamaguchi, T., Hosoya, H., and Gutman, I.,** *Tetrahedron,* 37, 1113, 1981.
43. **Clar, E.,** *Aromatic Sextet,* John Wiley & Sons, New York, 1972.
44. **Armitt, T. W. and Robinson, R.,** *J. Chem. Soc.,* 1604, 1925.
45. **Clar, E.,** *Polycyclic Hydrocarbons,* Part 1, Academic Press, London, 1964, 32.
46. **Randić, M.,** *Tetrahedron,* 31, 1477, 1975.
47. **Wheland, G. W.,** *J. Chem. Phys.,* 3, 356, 1935.
48. **Ohkami, N. and Hosoya, H.,** *Bull. Chem. Soc. Jpn.,* 52, 1624, 1979.

49. **Knop, J. V. and Trinajstić, N.,** *Int. J. Quantum Chem.,* S 14, 503, 1980.
50. **Kasum, D., Trinajstić, N., and Gutman, I.,** *Croat. Chem. Acta,* 54, 321, 1981.
51. **Sachs, H.,** *Publ. Math. (Debrecen),* 11, 119, 1964.
52. **Gordon, M. and Davison, W. H. T.,** *J. Chem. Phys.,* 20, 428, 1952.
53. **Yen, T. F.,** *Theor. Chim. Acta,* 20, 399, 1971; see also **Cvetković, D. and Gutman, I.,** *Croat. Chem. Acta.,* 46, 15, 1974.
54. **Platt, J. R.,** in *Handbuch der Physik,* Flugge, S., Ed., Springer-Verlag, Berlin, 1961, 173.
55. **Longuet-Higgins, H. C.,** *J. Chem. Phys.,* 18, 265, 1950; 18, 275, 1950; 18, 283, 1950.
56. **Dewar, M. J. S. and Dougherty, R. C.,** *The PMO Theory of Organic Chemistry,* Plenum Press, New York, 1975.
57. **Živković, T.,** *Croat. Chem. Acta,* 44, 351, 1972; see also **Graovac, A., Gutman, I., and Trinajstić, N.,** *Topological Approach to the Chemistry of Conjugated Molecules,* Lecture Notes in Chemistry, Vol. 4, Springer-Verlag, Berlin, 1977, 1.
58. **Randić, M.,** *Croat. Chem. Acta,* 47, 71, 1975.
59. **Randić, M., Ruščič, B., and Trinajstić, N.,** *Croat. Chem. Acta,* 54, 295, 1981.
60. **Eilfeld, P. and Schmidt, W.,** *J. Electron Spectr. Rel. Phenom.,* 24, 101, 1981.

Chapter 3

THE CONJUGATED CIRCUITS MODEL

I. CONJUGATED CIRCUITS

The conjugated circuits model for studying aromaticity and conjugation in polycyclic molecules has been introduced by Randić in a series of papers.[1-8] He carried out the graph theoretical analysis of Kekulé structures and found that each Kekulé structure can be partitioned in several conjugated circuits. Every circuit is represented by regular alternating of formally single and double bonds. Thus, conjugated circuits are necessarily of *even* length. The total number of circuits characterizes the system and provides a basis for discussion of the energetics of the conjugated structure.[1-14]

The circuit decomposition of individual Kekulé structures of polycyclic conjugated molecules leads to $(4m + 2)$ and/or $(4m)$; $m = 1, 2, \ldots$, (linearly independent, linearly dependent, and disconnected) conjugated circuits. Linearly independent circuits are those which cannot be represented as a superposition of conjugated circuits of smaller size. The linearly independent $(4m + 2)$ circuits are denoted by R_m, while $(4m)$ circuits by Q_m. For example, benzocyclobutadiene $(K = 3)$ consists of the following conjugated circuits,

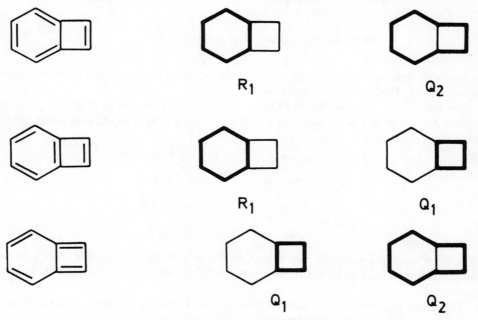

The total circuit decomposition of benzocyclobutadiene is $2R_1 + 2Q_1 + 2Q_2$.

The R_m and Q_m quantities may be used for defining the resonance energy of a conjugated molecule,

$$RE(\text{conjugated circuit}) = \frac{1}{K} \left\{ \sum_{m=1}^{N} (a_m R_m + b_m Q_m) \right\} \quad (1)$$

where a_m and b_m represent the number of R_m and Q_m circuits, respectively, for a given value of m. The advantage of relation (1) over many theoretical aromaticity indexes in literature is in that it reveals for the first time that RE indexes of various conjugated molecules are related among themselves through the parameters R_m and Q_m. Thus, the REs obtained by

Table 1
THE CIRCUITS DECOMPOSITIONS AND
RESONANCE ENERGIES OF SOME CONJUGATED
HYDROCARBONS

Molecule[a]	Circuits decomposition	Resonance energy (eV)
1	$1/2(2R_2)$	0.869
2	$1/3(4R_1 + 2R_2)$	1.323
3	$1/4(6R_1 + 4R_2 + 2R_3)$	1.600
4	$1/5(10R_1 + 4R_2 + R_3)$	1.955
5	$1/6(12R_1 + 8R_2 + 4R_3)$	2.133
6	$1/8(20R_1 + 10R_2 + 2R_3)$	2.505
7	$1/5(8R_1 + 2Q_1 + 4Q_2 + Q_3)$	0.360
8	$1/4(4R_1 + 2R_2 + 2Q_1 + 2Q_2 + 2Q_3)$	−0.108
9	$1/5(6R_1 + 2R_2 + 4Q_1 + 2Q_2 + Q_3)$	−0.349
10	$1/2(2R_2)$	0.246
11	$1/4(4R_1 + 2R_2 + 6R_3)$	1.142
12	$1/3(4R_1 + 2R_2)$	1.323
13	$1/3(4R_2 + 2Q_3)$	0.228
14	$1/2(2Q_2)$	−0.450
15	$1/4(4R_1 + 2Q_2 + 4Q_3 + 2Q_4)$	0.491
16	$1/3(4R_2 + 2Q_2)$	0.028
17	$1/4(8R_2 + 2R_3 + 2Q_3)$	0.475

[a]   Graphs of studied molecules are given in Figure 1.

means of the parameters R_m and Q_m may be used for a novel classification of the conjugated hydrocarbons. This will be discussed below.

The numerical values of R_m and Q_m are obtained from the parametrization procedure against the Dewar SCF MO resonance energies.[15] Values of R_m and Q_m parameters used by Randić are given below:

$$R_1 = 0.869 \text{ eV} \qquad Q_1 = -1.60 \text{ eV}$$

$$R_2 = 0.246 \text{ eV} \qquad Q_2 = -0.45 \text{ eV}$$

$$R_3 = 0.100 \text{ eV} \qquad Q_3 = -0.15 \text{ eV}$$

$$R_4 = 0.041 \text{ eV} \qquad Q_4 = -0.006 \text{ eV}$$

It is not necessary to carry the parametrization further to higher values of m $(m>4)$, because the correction of RE will be negligible due to small values of R_5, R_6, . . . and Q_5, Q_6, . . . parameters.

The above parameters, for example, produce the following resonance energy of benzo-cyclobutadiene (see the example above),

$$\text{RE(benzocyclobutadiene)} = \frac{3}{2}(R_1 + Q_1 + Q_2) = -0.787 \text{ eV}$$

Thus, benzocyclobutadiene is predicted to be an antiaromatic molecule, a result which agrees with the experimental observation[16] and other theoretical predictions.[17,18]

In Table 1, we give the circuit decompositions and resonance energies of a number of conjugated molecules.

The application of the conjugated circuits model produced several interesting results:

(i)   In benzenoid hydrocarbons, only conjugated circuits of size $(4m + 2) = R_m$ appear.
In Table 2, we give several benzenoids, their circuits' decompositions and resonance

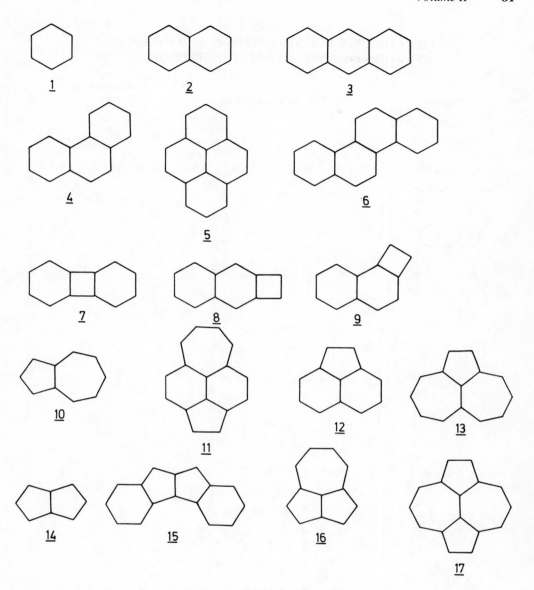

FIGURE 1.   Graphs of molecules in Table 1.

energies. Resonance energies of benzenoid hydrocarbons obtained by the use of the conjugated circuits model are on average within 0.05 eV of the values obtained by the SCF MO approach of Dewar[15] (see Table 3).

(ii)   Number of conjugated circuits in the linear polyacenes is dependent on the number of the rings, **R**, in a molecule. Linear polyacene may be schematically represented as below,

The number of hexagons in the linear polyacenes is denoted by **R**. Then, the following relations give the number of conjugated circuits R_m ($m = 1,2,3,4$) in terms of the number of rings in a given linear polyacene,

Table 2
THE CIRCUITS DECOMPOSITIONS AND RESONANCE
ENERGIES OF SOME BENZENOID HYDROCARBONS

Molecule	Circuits decomposition	Resonance energy (eV)
	$(10R_1 + 8R_2 + 6R_3 + 4R_4 + 12R_5)/6$	1.88
	$(22R_1 + 12R_2 + 7R_3 + 3R_4 + R_5)/9$	2.54
	$(36R_1 + 16R_2 + 6R_3 + 2R_4)/12$	2.99
	$(30R_1 + 18R_2 + 6R_3 + R_4)/11$	2.83
	$(40R_1 + 20R_2 + 5R_3)/13$	3.08
	$(42R_1 + 14R_2 + 5R_3 + 3R_4 + R_5)/13$	3.12
	$(40R_1 + 20R_2 + 5R_3)/13$	3.09
	$(26R_1 + 16R_2 + 5R_3 + 2R_4)/10$	2.71
	$(46R_1 + 18R_2 + 5R_3 + R_4)/14$	3.21
	$(42R_1 + 26R_2 + 12R_3 + 4R_4)/14$	3.16

Table 2 (continued)
THE CIRCUITS DECOMPOSITIONS AND RESONANCE ENERGIES OF SOME BENZENOID HYDROCARBONS

Molecule	Circuits decomposition	Resonance energy (eV)
	$(64R_1 + 48R_2 + 27R_3 + R_4)/20$	3.51
	$(60R_1 + 48R_2 + 36R_3 + 24R_4)/20$	3.43
	$(200R_1 + 160R_2 + 108R_3 + 32R_4)/50$	4.51

$$R_1 = 2 \cdot R$$

$$R_2 = R_1 - 2 = 2(R - 1)$$

$$R_3 = R_2 - 2 = 2(R - 2)$$

$$R_4 = R_3 - 2 = 2(R - 3)$$

- - - - - - - - - - - - - - - - - - - -

$$R_m = 2[R - (m - 1)]$$

Example

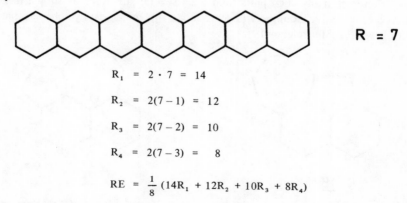

$$R = 7$$

$$R_1 = 2 \cdot 7 = 14$$

$$R_2 = 2(7 - 1) = 12$$

$$R_3 = 2(7 - 2) = 10$$

$$R_4 = 2(7 - 3) = 8$$

$$RE = \frac{1}{8}(14R_1 + 12R_2 + 10R_3 + 8R_4)$$

(iii) Decomposition of conjugation contained in Kekulé structures of nonalternant hydrocarbons reveals that they partition into *two* classes of compounds: one class is in which only conjugated circuits of $(4m + 2)$-type arise and the other contains structures with both, $(4m + 2)$ and $(4m)$, kinds of conjugated circuits. For example, the four Kekulé structures of acepleiadylene (**11**) give only R_m conjugated circuits

Table 3

COMPARISON OF THE RESONANCE ENERGIES (IN eV) OBTAINED FROM THE CONJUGATED CIRCUITS APPROACH AND FROM THE SCF MO APPROACH BY DEWAR

Molecule	Resonance energies (in eV)		Difference (in eV)
	Conjugated circuits	SCF MO	
Phenanthrene	1.96	1.93	0.03
Pyrene	2.13	2.10	0.03
Perylene	2.65	2.62	0.03
1.2-Benz-pyrene	2.61	2.58	0.03
4.5-Benz-pyrene	2.93	2.85	0.08
1.12-Benz-perylene	3.19	3.13	0.06
Chrysene	2.51	2.48	0.03
Triphenylene	2.72	2.65	0.07
Benz-anthracene	2.33	2.29	0.04
1.2,5.6- Dibenzanthracene	3.00	2.95	0.05

Average difference: 0.05

$(4R_1 + 2R_2 + 6R_3)$, although **11** is a nonalternant structure. In contrast another nonalternant azupyrene (**17**) gives $8R_2 + 2R_3 + 2Q_3$. Therefore, in analogy with the classification of alternant into benzenoids and nonbenzenoids,[19] the nonalternant systems may be classified into azulenoids and nonazulenoids.[3] The label *azulenoid* is used in view that azulene is one of the simplest and most commonly encountered nonalternant structure.[20] Examples of azulenoids and nonazulenoids are given in Figures 2 and 3.

(iv)   A novel feature emerging from the application of conjugated circuits model is the realization that different conjugated molecules may have *identical* decomposition of conjugation and, thus, identical resonance energies. Therefore, molecules with identical circuit decomposition, called *isoconjugated molecules*,[3] point to similar aromatic behavior.

There are many such pairs (and even several structures as shown in Figure 4). As an illustrative example will serve a pair of azulenoids: cycloheptacenaphthylene (**18**) and naphthazulene (**19**).

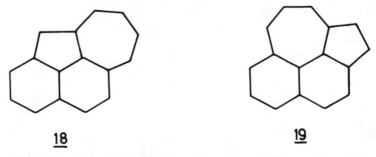

18   **19**

They have identical circuit decomposition: $4R_1 + 4R_2 + 4R_3$. Other theoretical methods for studying the aromatic stability also predict **18** and **19** to have similar aromatic stability, because they produced identical values of the aromaticity indexes for them.[3,21,22] However, these other methods could not trace the origin of this identity.

(v)   Randić[2] proposed *Aromaticity postulate:* ''Systems which possess *only* $(4m + 2)$

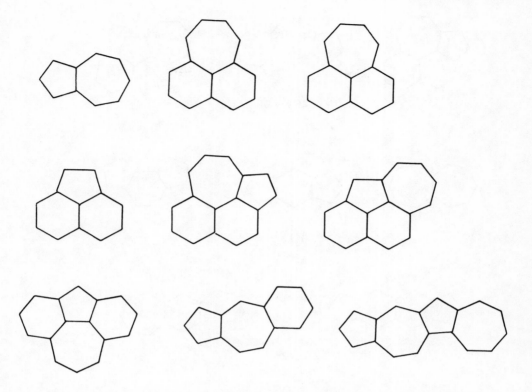

FIGURE 2.   Graphs of azulenoids (nonalternant molecules containing only $(4m + 2)$ conjugated circuits).

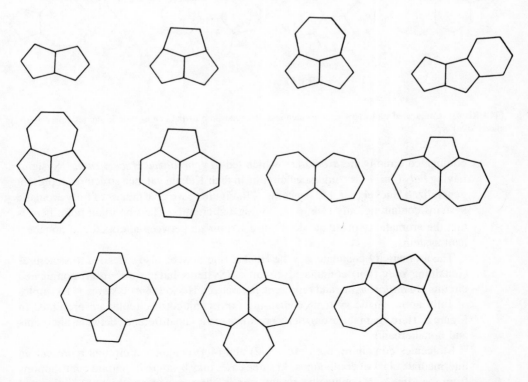

FIGURE 3.   Graphs of nonazulenoids (nonalternant molecules containing $(4m + 2)$ and/or $(4m)$ conjugated circuits).

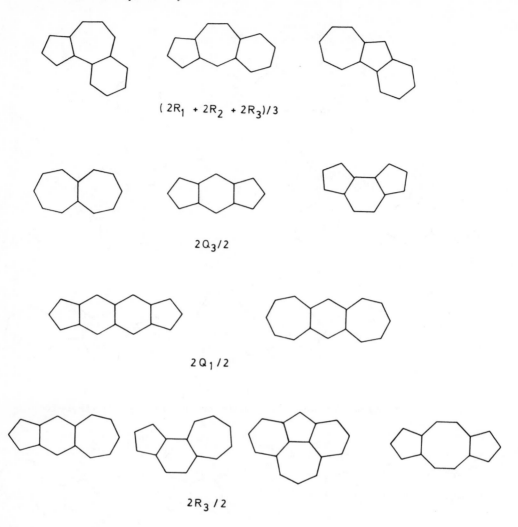

$$(2R_1 + 2R_2 + 2R_3)/3$$

$$2Q_3/2$$

$$2Q_1/2$$

$$2R_3/2$$

FIGURE 4. Conjugated molecules with an identical decomposition of the conjugation in circuits of $(4m + 2)$ and/or $(4m)$ type

conjugated circuits are *aromatic*'' which led to *generalized Hückel rule:* ''Systems having *only* $(4m + 2)$ conjugated circuits in their Kekulé valence structures represent generalized Hückel $(4m + 2)$ systems.'' In Figure 5, we give graphs of fully aromatic systems containing only $(4m + 2)$ conjugated circuits. The important point here is that the aromaticity postulate does not discriminate between alternant and nonalternant systems.

The aromaticity postulate and the Hückel rule may be also extended to structures containing *only* $(4m)$ conjugated circuits: ''Systems having *only* $(4m)$ conjugated circuits are *antiaromatic* and represent generalized Hückel $(4m)$ systems.'' Examples of fully antiaromatic systems containing $(4m)$ conjugated circuits are presented in Figure 6. Here again the aromaticity postulate makes no difference between alternants and nonalternants.

Molecules containing both $(4m + 2)$ and $(4m)$ conjugated circuits represent an intermediate class of compounds. In some cases may be more important contributions from the $(4m + 2)$ conjugated circuits and in other cases from the $(4m)$ conjugated

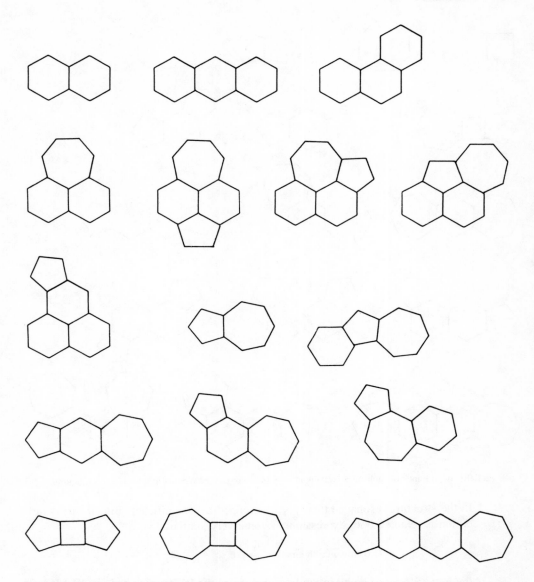

FIGURE 5.   Graphs of molecules containing only $(4m + 2)$ conjugated circuits in their Kekulé structures.

circuits. The former systems exhibit features similar to aromatic structures, while the latter systems resemble antiaromatic species. Examples of both classes of molecules are given in Figures 7 and 8, respectively.

(vi)   The conjugated circuits approach is related to the structure-resonance theory,[23] although this does not appear as such at first sight. However, the closeness of the two methods is perceptible in actual application.[24] Let us consider, for example, phenanthrene by both methods. The conjugated circuits method of Randić produces for the resonance energy of phenanthrene.[1]

$$RE(\text{conjugated circuits}) \;=\; \frac{1}{5}\,(10R_1 + 4R_2 + R_3)$$

while the Herndon's structure-resonance theory gives,[23]

$$RE(\text{resonance energy}) \;=\; \frac{1}{5}\,(10_{\gamma_1} + 4_{\gamma_2})$$

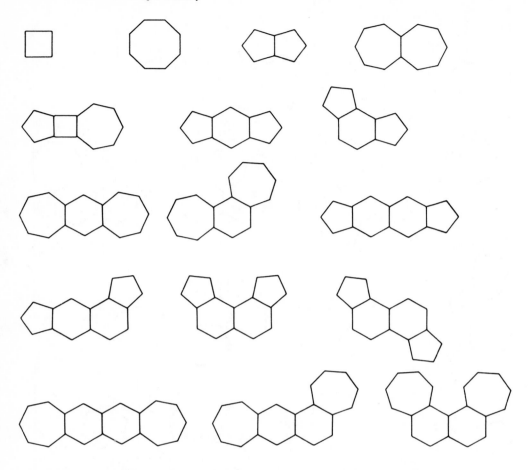

FIGURE 6. Graphs of molecules containing only (4*m*) conjugated circuits in their Kekulé structures.

In the structure-resonance theory γ_3 is dropped as insignificant. But, if this contribution is retained both expressions become very similar,

$$RE(\text{resonance theory}) \;=\; \frac{1}{5}\left(10_{\gamma_1} + 4_{\gamma_2} + 2_{\gamma_3}\right)$$

the only difference being in the last term, since the RE (conjugated circuits) is built up from the linearly independent circuits only. However, if the linearly dependent circuits were included in the model, the two methods would become identical. Thus, the Randić's resonance energy for phenanthrene with inclusion of all circuits becomes,

$$RE(\text{conjugated circuits}) \;=\; \frac{1}{5}\left(10R_1 + 4R_2 + 2R_3\right)$$

The meaning of this result is that the Randić's conjugated circuits method and the Herndon's structure-resonance theory become identical if the Randić's rule for rejecting linearly dependent circuits is discarded and if all γ_i (or ω_i; i = 1,2,3, and 4) integrals are kept in the Herndon's theory.

Let us end this section by mentioning that the conjugated circuits model may be generalized.[12,25] Apparently, Randić[26] has succeeded in doing just that, thus, extending the power and the range of applicability of the model. For example, Randić has applied the conjugated circuits model to dianions of conjugated hydrocarbons,[27] Möbius structures[27] and heterocycles.[28]

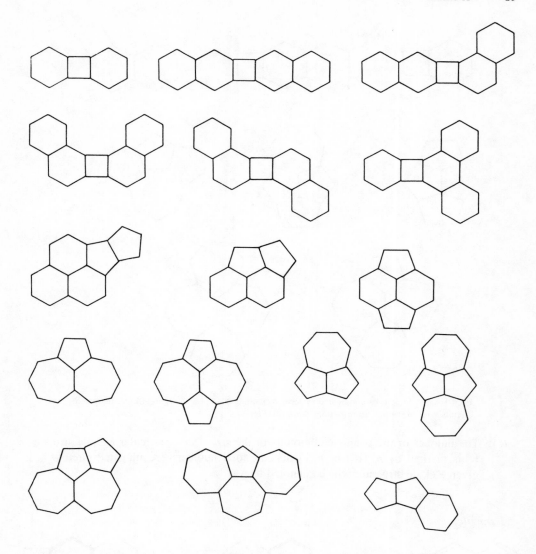

FIGURE 7.   Graphs of conjugated systems containing both $(4m + 2)$ and $(4m)$ conjugated circuits with prevailing contribution from $(4m + 2)$ circuits.

II. SOME PROPERTIES OF CONJUGATED CIRCUITS

In this section, several properties of conjugated circuits will be discussed. To make the discussion simpler we need the following notation:

(a)   Conjugated circuits of the size a will be denoted by C_a ($a = 4,6,8,10, \ldots$) Note, in the further discussion we will use the notation $C_4 \equiv Q_1$, $C_6 \equiv R_1$, $C_8 \equiv Q_2, \ldots$

(b)   The j-th Kekulé structure of a given conjugated molecule will be symbolized by K_j. Note that

$$K = \sum_j K_j$$

where K is the total number of Kekulé structures in a molecule.

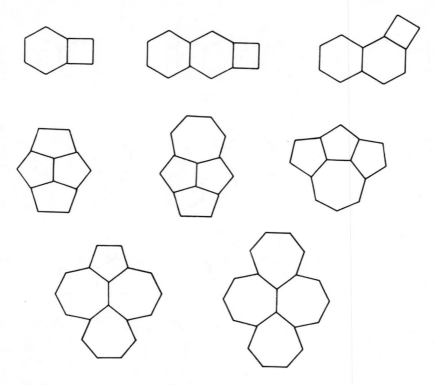

FIGURE 8.   Graphs of conjugated systems containing both $(4m + 2)$ and $(4m)$ conjugated circuits with prevailing contributions from $(4m)$ circuits.

(c)   The number of the conjugated circuits of the size a in a particular Kekulé structure K_j is marked by $n_a (K_j)$ or n_a. Thus, the total number of conjugated circuits in a given Kekulé structure will be denoted by

$$\sum_a n_a C_a.$$

Example

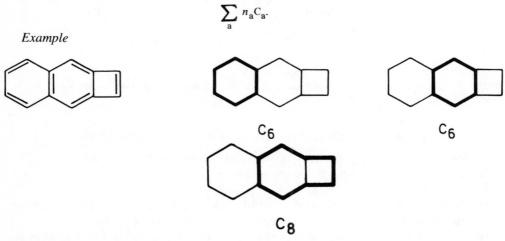

Decomposition expression: $2 C_6 + C_8$

(d)   A pair of mutually disjoint conjugated circuits with respective sizes a and b will be denoted by C_{ab}. Similarly, triplets, quartets, . . . , etc. of mutually disconnected conjugated circuits will be symbolized by $C_{a,b,c}$, $C_{a,b,c,d}$, . . . , etc. The number of such circuits in a given Kekulé structure K_j is denoted by $n_{a,b}(K_j)$, $n_{a,b,c}(K_j)$, $n_{a,b,c,d}(K_j)$, . . . , etc. or $n_{a,b}$, $n_{a,b,c}$, $n_{a,b,c,d}$, . . . , etc.

Below we list some properties of conjugated circuits.[6,7]
(1)   If K = 1, then n_a = 0 for all values of *a*.
(2)   If K > 1, then n_a > 0 at least for some *a*.
(3)   The benzenoid systems contain only conjugated circuits of size (4m + 2), i.e., n_a = 0 whenever a ≡ 0 (mod 4). If a benzenoid molecule contains no essential single bonds, then the number of linearly independent conjugated circuits in each Kekulé structure is the same and equals the number of fused rings in a molecule.

Example

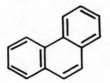

Number of rings: 3
Number of expected linearly independent conjugated circuits: 3

phenanthrene

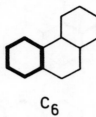

C_6 C_{10} C_6

If, however, there are present essential single bonds, in a molecule, a reduced system, obtained after removing all essential single bonds, should be considered. The number of linearly independent conjugated circuits is then equal to the number of fused rings in the reduced system.

Example

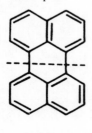

Number of rings: 4
Number of expected linearly independent conjugated circuits: 4

pyrene   **reduced system**

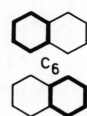

C_6 C_6

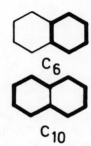

C_6 C_{10}

(4)   The number of all conjugated circuits (including linearly dependent and disjoint conjugated circuits) in each Kekulé structure is same and equals K − 1,

$$\sum_a n_a(K_j) + \sum_{a,b} n_{a,b}(K_j) + \sum_{a,b,c} n_{a,b,c}(K_j) + \ldots = K - 1 \qquad (2)$$

Therefore, by examining an arbitrary Kekulé structure, the number of all Kekulé structures of a molecule may be obtained.

Example

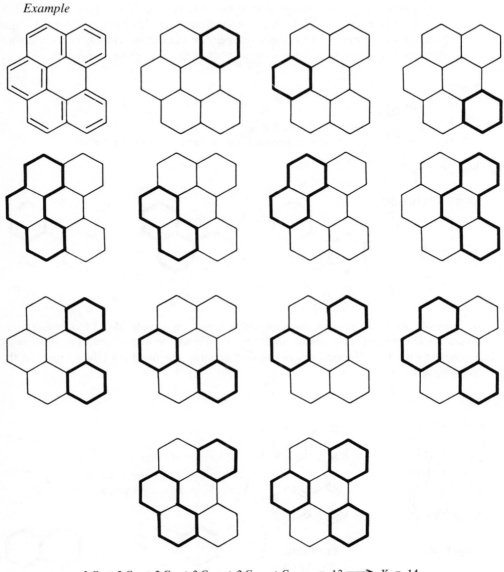

$$3\,C_6 + 2\,C_{10} + 2\,C_{14} + 3\,C_{6,6} + 2\,C_{6,10} + C_{6,6,6} = 13 \Longrightarrow K = 14$$

III. THE CIRCUIT POLYNOMIAL

The circuit polynomial[29,30] is an important combinatorial structure in the conjugated circuits model of Randić, because it may be used for counting circuits in the conjugated molecules. Circuit polynomials for several benzenoid hydrocarbons have been reported by Gutman and Randić.[6] However, they did not define the circuit polynomial.

Let G be a graph of an arbitrary conjugated polycyclic molecule with K $\neq$ 0. Let C_ℓ denote the conjugated circuits of G of a length ℓ, and let S be the maximal number so that C_{2S+2} exists in G. Let L be the maximal number of mutually disconnected circuits in G, i.e., there exist L disconnected circuits C_{ℓ_i}, $i = 1, 2, \ldots, L$; ℓ_i being appropriate integers.

Let for a given k (running from 1 to L and denoting the number of disconnected circuits) indicate,

$$M(k) = \left\{ \left(C_{\ell_i}, \ldots, C_{\ell_k} \right) \middle| C_{\ell_i} \text{ being mutually disconnected circuits} \right\}$$

(3)

If $m(k) = (C_{\ell_1}, \ldots, C_{\ell_k})$ is an element of $M(k)$, then we can symbolize by $p(m[k])$ the number of Kekulé structures of the subgraphs

$$G - \bigcup_{i=1}^{k} C_{\ell_i}$$

In the case

$$G = \bigcup_{i=1}^{k} C_{\ell_i}$$

we define $p(m[k]) = 1$. If, for the sake of convenience, we denote,

$$m_i = \frac{\ell_i - 2}{2}$$

(4)

we can define the *circuit (or cyclic) polynomial* $P^c(G; y_1, y_2, \ldots, y_S)$ by,[30]

$$P^c(G; y_1, y_2, \ldots, y_S) = K(G) + \sum_{k=1}^{L} \sum_{m(k) \in M(k)} p(m[k]) \prod_{i=1}^{k} y_{m_i}$$

(5)

where $K(G) = K$. $m(k) = (C_{\ell_1}, \ldots, C_{\ell_k})$ and $m(k) = C_{f_1}, \ldots, C_{f_k})$ are equivalent if (after the appropriate ordering) $\ell_i = f_i$ for $i = 1, 2, \ldots, k$. Then we can define *circuit numbers* of the graph G, denoted by $p(\ell_1, \ell_2, \ldots, \ell_k)$, by,

$$p(\ell_1, \ell_2, \ldots, \ell_k) = \sum_{m(k) \in M_i(k)} p(m[k])$$

(6)

where $M_i(k)$ is the class of equivalence to which element $m(k)$ belongs. The sum of all circuit numbers gives the number of all conjugated circuits, including linearly dependent and disjoint circuits.

Example

The construction of the circuit polynomial of G

G

The procedure involves several steps.

(A)   *The enumeration of Kekulé structures*

The number of Kekulé structures may be obtained by one of the methods described in the previous chapter. For the fast and reliable enumeration of Kekulé valence forms of the complex structures, including fragments of the type

$$G - \bigcup_{i=1}^{k} C_{\ell_i}$$

computer program has been recently made available.[31] The above molecule has five Kekulé structures:

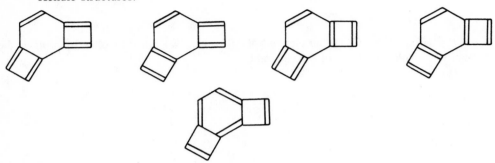

(B) *The evaluation of the circuit numbers of G*

(B.1) k = 1

$\ell_1 = 4, m_1 = 1$

$p(4) =$ 2   +   2   = 4

$\ell_1 = 6, m_1 = 2$

$p(6) =$ 1

$\ell_1 = 8, m_1 = 3$

$p(8) =$ 1   +   1   = 2

$\ell_1 = 10, m_1 = 4$

$p(10) =$ 1

(B.2) k = 2

$\ell_1 = 4, m_1 = 1$

$\ell_2 = 4, m_2 = 1$

$p(4,4) =$

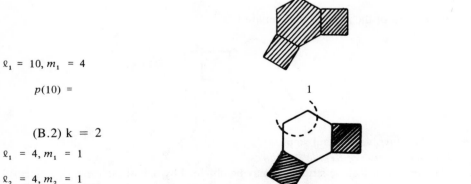

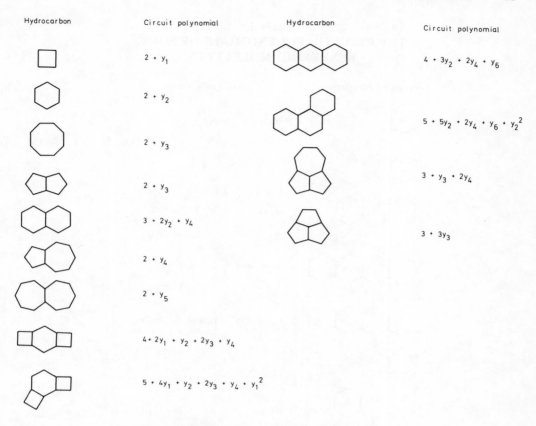

FIGURE 9. Circuit polynomials of some conjugated hydrocarbons.

(C) *The circuit polynomial of G*
The circuit polynomial of G is then given by,

$$P^c(G; y_1, y_2, y_3, y_4) = 5 + 4y_1 + y_2 + 2y_3 + y_4 + y_1^2$$

Circuit polynomials of some conjugated hydrocarbons are given in Figure 9.

A. Some Properties of the Circuit Polynomial

The circuit polynomial has some interesting properties. Several of them are listed below.

1. Since the benzenoid hydrocarbons contain only $(4m + 2)$ conjugated circuits, the polynomial (5) may be rewritten to the following by substituting in (5) for $y_{2m} = x_m$ and $y_2 m_{-1} = 0$,

$$P^c(G; x_1, x_2, \ldots, x_R) = K(G) + \sum_{k=1}^{L} \sum_{m(k) \in M(k)} p(m[k]) \prod_{i=1}^{k} x_{m_i}$$

(7)

where R is a number of hexagonal rings. (See also Table 4.)

Table 4
THE CIRCUIT   POLYNOMIALS OF SOME
BENZENOID MOLECULES

Benzenoid hydrocarbon	Circuit polynomial
	$P^C(G; x) = 2 + x_1$
	$P^C(G; x) = 3 + 2x_1 + x_2$
	$P^C(G; x) = 4 + 3x_1 + 2x_2 + x_3$
	$P^C(G; x) = 5 + 4x_1 + 3x_2 + 2x_3 + x_4$
	$P^C(G; x) = 8 + 10x_1 + 5x_2 + 2x_3 + x_4$
	$P^C(G; x) = 7 + 8x_1 + 4x_2 + 2x_3 + x_4$
	$P^C(G; x) = 9 + 13x_1 + 3x_2 + 3x_3 + x_4$
	$P^C(G; x) = 6 + 6x_1 + 4x_2 + 2x_3$

Example

The construction of the circuit polynomial for 1,2-benzanthracene by means of Equation (7)

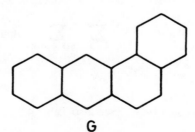

G

(A)   *Enumeration of Kekulé structures*

In this case the most convenient method for the enumeration of Kekulé structures is the Gordon-Davison procedure[32] which gives straightforwardly $K(G) = 7$.

(B)   *The evaluation of the circuit numbers of G*

(B.1) $k = 1$

$\ell_1 = 6, m_1 = 1$

$p(6) = \quad\quad 2 \quad + \quad 2 \quad + \quad 1 \quad + \quad 3 \quad = 8$

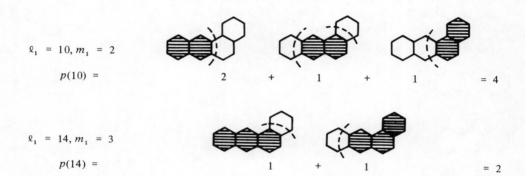

$\ell_1 = 10, m_1 = 2$

$p(10) = \quad\quad 2 \quad + \quad 1 \quad + \quad 1 \quad\quad = 4$

$\ell_1 = 14, m_1 = 3$

$p(14) = \quad\quad 1 \quad + \quad 1 \quad\quad = 2$

$\ell_1 = 18, m_1 = 4$

$p(18) = \quad\quad 1$

(B.2) $k = 2$

$\ell_1 = 6, m_1 = 1$

$\ell_2 = 6, m_2 = 1$

$p(6,6) = \quad\quad 1 \quad + \quad 0 \quad + \quad 1 \quad\quad = 2$

$\ell_1 = 6, m_1 = 1$

$\ell_2 = 10, m_2 = 2$

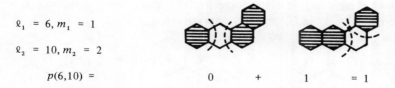

$p(6,10) = \quad\quad 0 \quad + \quad 1 \quad\quad = 1$

(C)   *The circuit polynomial of 1,2-benzanthracene*

$$P^c(G; x_1, x_2, x_3, x_4) = 7 + 8x_1 + 4x_2 + 2x_3 + x_4 + 2x_1^2 + x_1 x_2$$

The circuit polynomials of several benzenoid hydrocarbons are given in Figure 10.

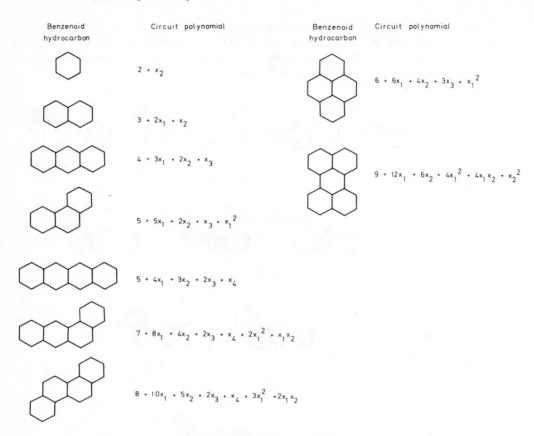

FIGURE 10.   Circuit polynomials of several benzenoid hydrocarbons.

2.   In the case of linear polyacenes the circuit polynomial may be defined only in terms of the number of hexagonal rings. Let G_r be a graph representing a linear polyacene with r six-membered rings.

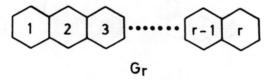

Let also graphs G_{r_1} and G_{r_2} denote two molecules: one with a single Kekulé structure (G_{r_1}) and the other with no Kekulé structures (G_{r_2}), respectively. r_1 and r_2 stand, respectively, for a number of six-membered rings in G_{r_1} and G_{r_2}, respectively.

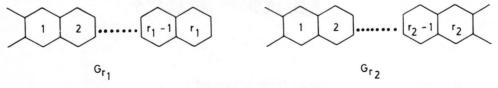

where r_1 and r_2 are, respectively, the numbers of hexagonal rings in G_{r_1} and G_{r_2}. It is easy to show that the molecule represented by the graph G_n has only *one* Kekulé structure for each r_1, while structure G_{r_2} has *no* Kekulé structures for each value of r_2, i.e., $K(G_{r_1}) = 1$ and $K(G_{r_2}) = 0$.

Since graph $G\text{-}C_{4m_1+2}$ for $k = 1$ is either graph G_{r_1} or a union of two graphs of the type G_{r_1}, it is evident that,

$$p(6) \ = r$$

$$p(10) = r - 1$$

$$p(14) = r - 2$$

$$\cdots\cdots\cdots$$

$$p(4r - 2) = 2$$

$$p(4r + 2) = 1 \text{ (by definition)} \tag{8}$$

Because

$$G - \bigcup_{i=1}^{k} C_{\varrho_i}$$

for $k \geq 2$ is either a graph G_{r_2} or a union of graphs G_{r_1} and G_{r_2}, it is clear that

$$p\left(G - \bigcup_{i=1}^{k} C_{\varrho_i}; k\right) = 0$$

for every $k \geq 2$. Then, the cyclic polynomial of a linear polyacene has the following form,

$$P^c(G_r, x_1, x_2, \ldots, x_r) = (r + 1) + rx_1 + (r - 1)x_2 + \ldots$$

$$\ldots + 2x_{r-1} + x_r \tag{9}$$

Example

The circuit polynomial of pentacene

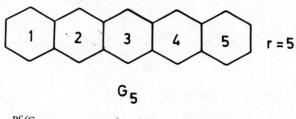

$$G_5$$

$$P^c(G_r; x_1, x_2, x_3, x_4, x_5) = 6 + 5x_1 + 4x_2 + 3x_3 + 2x_4 + x_5$$

3.   The circuit (cyclic) polynomial, as defined here, has a useful combinatorial property. If we formally substitute for $y_{m_i} = 2C_{2m_i+2}$ $(m_i = 1, 2, \ldots, S)$ in Equation (5) or for $x_{m_i} = 2C_{4m_i+2}$ $(m_i = 1, 2, \ldots, R)$ in (7), and delete $K(G)$, we obtain expressions for counting circuits in a given graph.

$$\sum_{k=1}^{L} \sum_{m(k) \in M(k)} p(m[k]) \, 2^k \prod_{i=1}^{k} C_{2m_i + 2} \tag{10}$$

or for enumeration of conjugated circuits in benzenoids,

$$\sum_{k=1}^{L} \sum_{m(k) \in M(k)} p(m[k]) \, 2^k \prod_{i=1}^{k} C_{4m_i + 2} \tag{11}$$

Example

Counting of conjugated circuits in biphenylene

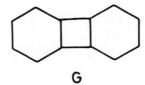

G

- The cyclic polynomial of biphenylene

$$P^c(G; y_1, y_2, y_3, y_5) = 5 + y_1 + 4y_2 + 2y_3 + y_5 + y_2^2$$

- Substitution of y_{m_i} for $2C_{2m_i+2}$ in the cyclic polynomial

$$y_1 = 2C_4$$

$$y_2 = 2C_6$$

$$y_3 = 2C_8$$

$$y_5 = 2C_{12}$$

- The circuit count for biphenylene

$$2C_4 + 8C_6 + 4C_8 + 2C_{12} + C_6 \cdot C_6$$

Counting of conjugated circuits in pentacene

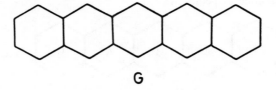

G

- The cyclic polynomial of pentacene

$$P^c(G; x_1, x_2, x_3, x_4, x_5) = 6 + 5x_1 + 4x_2 + 3x_3 + 2x_4 + x_5$$

- Substitution of x_{m_i} for $2C_{4m_i+2}$ in the circuit polynomial

$$x_1 = 2C_6$$

$$x_2 = 2C_{10}$$

$$x_3 = 2C_{14}$$

$$x_4 = 2C_{18}$$

$$x_5 = 2C_{22}$$

- The circuit count for pentacene

$$10C_6 + 8C_{10} + 6C_{14} + 4C_{18} + 2C_{22}$$

The expression for counting the conjugated circuits of linear polyacenes becomes very simple after substituting $x_{m_i} = 2C_{4m_i+2}$ ($m_i = 1, 2, \ldots, r$) in Equation 9),

$$2(r + 1 - m_i)\, C_{4m_i + 2} \tag{12}$$

IV. PARITY OF CONJUGATED CIRCUITS

So far it was shown that the concept of conjugated circuits was very useful for expressing resonance energy, for defining aromaticity, for classifying conjugated systems, etc. In this section, we wish to clarify some ambiguities concerning the parity of Kekulé structures by means of conjugated circuits.

In Section II. A of Chapter 2, we have seen that the parity concept breaks down for some odd-membered polycyclic hydrocarbons with three or more cycles. In addition, it appears that the assumption that the Kekulé structures of the same parity make the same kind of contributions to the stability of a molecule is also questionable. However, the concept of parity may be generalized by applying it to conjugated circuits rather than to Kekulé structures.[4] Thus, all $(4m + 2)$ conjugated circuits are given positive parity and conjugated circuits of $(4m)$ size are given negative parity. This then permits assigning to each Kekulé structure an integer number, called ACCC (algebraic conjugated circuits count) which gives the difference of the number of conjugated circuits of $(4m + 2)$ and $(4m)$ kind. The ACCC label is introduced in parallel to ASC (algebraic structure count)[33] which is the difference between the number of Kekulé structures of opposing parity. A closer examination of the proposed scheme shows a full agreement in many structures for which the concept of parity operates, while ambiguities in other instances have been removed. An important conceptual distinction, however, is introduced: the ACCC values are *absolute* parameters not *relative* as before. This is an advantage, because the new label can be associated with individual structural formulae. In case of previously introduced parity, concept one would find that symmetry equivalent structures may have opposite parity and make opposite contribution to resonance energy; e.g., two Kekulé structures of cyclobutadiene. Such difficulties vanish when the ACCC parameters are used, since they can take any positive, zero, or negative values, in contrast to $+1$ and -1 values for the relative parities. Thus, for example, in cyclobutadiene both structures have ACCC $= -1$, they are both negative, and the overall sum of ACCC is suggestive that the molecule is antiaromatic. In general, valence structures with positive ACCCC will play a stabilizing role. Structures with zero ACCC may be called nonstabilizing, while those with negative ACCC will be destabilizing. In Figure 11, the ACCC values of biphenylene are given. In the case of biphenylene the relative parities and the ACCC quantities parallel each other. However, there are cases where the incompatibility between them occurs.

Example

$$2\,Q_1 + 4\,Q_2$$

$$ACCC = -6$$

$$p = +1$$

In some cases all Kekulé structures have the same ACCC value.

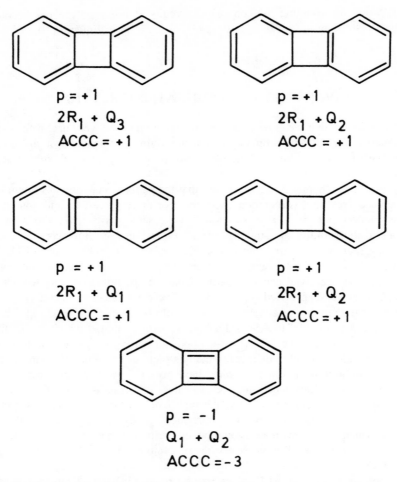

FIGURE 11.   The parity assignment, the circuit decomposition, and the ACCC values of Kekulé structures of biphenylene.

Example

Acepentylene

ACCC =-2

Benzodicyclobutadiene

ACCC =-1

Dibenzobutylene

ACCC = 0

In dibenzobutylene and benzodicyclobutadiene, half of the Kekulé structures have positive parity and half have negative parity, while in the ACCC scheme all the Kekulé structures belong to the same class of nonstabilizing or destabilizing structures, respectively. The overall conclusion is in both cases similar: a rather unstable system. For acepentylene the parity criterion is ambiguous: a tacitly assumed compatibility of the parities derived from pairs of structures does not hold.[34,35] However, there are no problems with the ACCC parameters: they are assigned to individual Kekulé structures, and since these are all equivalent, they have all same ACCC values of -2.

REFERENCES

1. **Randić, M.,** *Chem. Phys. Lett.,* 38, 68, 1976.
2. **Randić, M.,** *J. Am. Chem. Soc.,* 99, 444, 1977.
3. **Randić, M.,** *Tetrahedron,* 33, 1905, 1977.
4. **Randić, M.,** *Mol. Phys.,* 34, 849, 1977.
5. **Graovac, A., Gutman, I., Randić, M., and Trinajstić, N.,** *Collect. Czech. Chem. Commun.,* 43, 1375, 1978.
6. **Gutman, I. and Randić, M.,** *Chem. Phys.,* 41, 265, 1979.
7. **Randić, M.,** *Int. J. Quantum Chem.,* 17, 549, 1980.
8. **Randić, M.,** *Pure Appl. Chem.,* 52, 1587, 1980.
9. **Gayoso, J.,** *C. R. Acad. Sci. Ser. C.,* 288, 327, 1979.
10. **Gomes, J. A. N. F. and Mallion, R. B.,** *Rev. Port. Quim.,* 21, 82, 1979.
11. **Gomes, J. A. N. F.,** *Croat. Chem. Acta,* 53, 561, 1980.
12. **Gomes, J. A. N. F.,** *Theor. Chim. Acta,* 59, 333, 1981.
13. **El-Basil, S.,** *Int. J. Quantum Chem.,* 19, 593, 1981.
14. **El-Basil, S.,** *Math. Chem. (Mülheim/Ruhr),* 11, 97, 1981.
15. **Dewar, M. J. S. and de Llano, C.,** *J. Am. Chem. Soc.,* 91, 789, 1969.
16. **Chapman, O. L., Chang, C. C., and Rosenquist, N. R.,** *J. Am. Chem. Soc.,* 98, 261, 1976.
17. **Hess, B. A., Jr. and Schaad, L. J.,** *J. Am. Chem. Soc.,* 93, 305, 1971.
18. **Gutman, I., Milun, M., and Trinajstić, N.,** *J. Am. Chem. Soc.,* 99, 1692, 1977.
19. **Coulson, C. A. and Longuet-Higgins, H. C.,** *Proc. R. Soc. London, Ser. A,* 192, 16, 1947.
20. **Heilbronner, E.,** in *Non-Benzenoid Aromatic Compounds,* Ginsburg, D., Ed., Wiley-Interscience, New York, 1959, 171.
21. **Hess, B. A., Jr. and Schaad, L. J.,** *J. Am. Chem. Soc.,* 94, 3068, 1972.
22. **Herndon, W. C. and Ellzey, M. L., Jr.,** *J. Am. Chem. Soc.,* 96, 6631, 1976.
23. **Herndon, W. C.,** *Isr. J. Chem.,* 20, 270, 1980.
24. **Schaad, L. J. and Hess, B. A., Jr.,** *Pure Appl. Chem.,* 54, 1097, 1982.
25. **El-Basil, S.,** private communication.
26. **Randić, M.,** private communication.
27. **Randić, M.,** *J. Phys. Chem.,* 86, 3970, 1982.
28. **Randić, M., Jeričević, Ž., and Trinajstić, N.,** in preparation.
29. **Knop, J. V. and Trinajstić, N.,** *Int. J. Quantum Chem.,* S 14, 503, 1980.
30. **Seibert, J. and Trinajstić, N.,** *Int. J. Quantum Chem.,* in press.
31. **Džonova-Jerman-Blažić, B. and Trinajstić, N.,** *Computers & Chemistry,* 6, 121, 1982.
32. **Gordon, M. and Davison, W. H. T.,** *J. Chem. Phys.,* 20, 428, 1952.
33. **Wilcox, C. F., Jr.,** *Tetrahedron Lett.,* 795, 1968.
34. **Gutman, I. and Trinajstić, N.,** *Croat. Chem. Acta,* 47, 35, 1975.
35. **Gutman, I., Randić, M., and Trinajstić, N.,** *Rev. Roum. Chim.,* 23, 383, 1978.

Chapter 4

TOPOLOGICAL INDEXES AND THEIR APPLICATIONS TO STRUCTURE-PROPERTY AND STRUCTURE-ACTIVITY RELATIONSHIPS

A graph theoretical characterization of molecules may be realized sometimes uniquely (by means of a matrix or by means of a sequence of numbers) and sometimes not uniquely (by means of a polynomial or by means of a numerical index). A numerical index characterizing a molecule is called *topological index*.[1,2] A topological index is, therefore, a numerical descriptor of a molecule, based on a certain topological feature of the corresponding graph. In other words, a topological index attempts to express numerically the topological information for a given chemical compound. Topological information usually gives a ''hint'' about the size and shape, i.e., the nature, of a chemical structure.

The advantage of topological indexes is in that that they may be used directly as simple numerical descriptors in a comparison with physical, chemical, or biological parameters of molecules in quantitative structure-property relationships (QSPR) and in quantitative structure-activity relationships (QSAR).[1,3,4]

I. GRAPH THEORETICAL FOUNDATIONS OF TOPOLOGICAL INDEXES

Most of the proposed topological indexes are related either to a vertex adjacency relationship (atom-atom connectivity) in the graph (molecular structure) G or to topological distances in G. Therefore, the origin of topological indexes can be traced either to the adjacency matrix of a graph or to the distance matrix of a graph.

A. Indexes Based on Connectivity

Topological indexes related to the atom-atom connectivity in G are based either on the total sum of some combination of degrees of the adjacent vertices or on the graph spectrum. Several of these will be described below roughly in the order of their appearances.

1. The Zagreb Group Indexes

In the early work of the Zagreb Group[5] on the topological basis of π-electron energy, two terms appeared in the approximate formula for the total π-energy of a conjugated system which may be used separately as topological indexes,[6]

$$M_1(G) = \sum_{j=1}^{N} D_j^2 \tag{1}$$

$$M_2(G) = \sum_{(i,j)} D_i D_j \tag{2}$$

The symbol D_i stands for the valency of a vertex i. The sum in (1) is over all vertices of G, while the sum in (2) is over all edges in G.

2. The Connectivity Index

The connectivity index of a graph, $\chi_R(G)$, is introduced by Randić[7] and it is similar to the Zagreb Group indexes. It was obtained after the search for the specific index to characterize the branching of acyclic saturated hydrocarbons, unlike the indexes $M_1(G)$ and $M_2(G)$ which appeared in the topological formula for the π-electron energy of conjugated systems. The connectivity index of a graph was defined by Randić as,

$$\chi_R(G) = \sum_{\text{edges}} (D_i D_j)^{-1/2} \tag{3}$$

Randić's connectivity index may be generalized by considering a path of length L instead of an edge (L = 1) in the graph,[8]

$$^L\chi_R(G) = \sum_{\text{paths}} (D_i D_j \dots D_{L+1})^{-1/2} \tag{4}$$

where $D_i, D_j, \dots, D_{L+1}$ are valencies of vertices in the considered path L. From (4) naturally follow the three connectivity indexes most often used:[3]

(i) The zero-order connectivity index

$$^0\chi_R(G) = \sum_{s=1}^{s_v} (D_i)_s^{-1/2} \tag{5}$$

where s stands for a subgraph of G which is in this case just a vertex, while s_v is the total number of vertices in G. Each vertex of G has in this case a weight D_i.

(ii) The first-order connectivity index

$$^1\chi_R(G) = \sum_{s=1}^{s_e} (D_i D_j)_s^{-1/2}; \ i \neq j \tag{6}$$

where s stands for an edge in G, while s_e is the total number of edges in G. Each edge of G has in this case a weight of $D_i D_j$. The first-order connectivity index is, of course, identical to the original Randić's connectivity index.

(iii) The second-order connectivity index

$$^2\chi_R(G) = \sum_{s=1}^{s_L} (D_i D_j D_k)_s^{-1/2}; \ i \neq j \neq k \tag{7}$$

where s stands for a path of length two, while s_L is the number of paths of length two in G. Each path of length two has in this case a weight $D_i D_j D_k$. Higher order connectivity indexes may also be obtained directly from (4) if needed.

The calculation of $^0\chi_R(G)$, $^1\chi_R(G)$, and $^2\chi_R(G)$ indexes will be illustrated below for isopentane.

Example

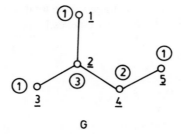

G

The underlined numbers represent the labeling of the graph G, while the encircled numbers stand for the valencies of the vertices in G.

(i) The zero-order connectivity index

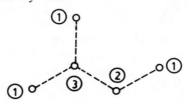

$$^0\chi_R(G) = 1 + (3)^{-1/2} + 1 + (2)^{-1/2} + 1 = 4.28$$

(ii) The first-order connectivity index

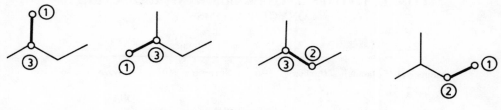

$$^1\chi_R(G) = 2(1\cdot3)^{-1/2} + (2\cdot3)^{-1/2} + (1\cdot2)^{-1/2} = 2.27$$

(iii) The second-order connectivity index

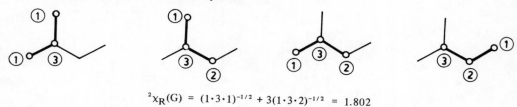

$$^2\chi_R(G) = (1\cdot3\cdot1)^{-1/2} + 3(1\cdot3\cdot2)^{-1/2} = 1.802$$

The relationship between the connectivity index and the adjacency matrix is rather simple. Since the adjacency matrix $A(G)$ has only entries 1 and 0, it expresses both the edge count and the edge location. Thus, $^L\chi_R(G)$ can be expressed in terms of the elements A_{ij},

$$^L\chi_R(G) = \sum_{i=1}^{N} \sum_{j>i}^{N} (D_i D_j \dots D_{L+1})^{-1/2} A_{ij}^L \tag{8}$$

Example

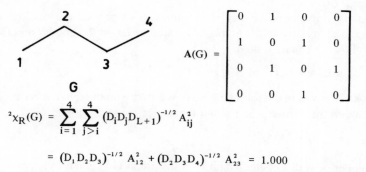

$$A(G) = \begin{bmatrix} 0 & 1 & 0 & 0 \\ 1 & 0 & 1 & 0 \\ 0 & 1 & 0 & 1 \\ 0 & 0 & 1 & 0 \end{bmatrix}$$

$$^2\chi_R(G) = \sum_{i=1}^{4} \sum_{j>i}^{4} (D_i D_j D_{L+1})^{-1/2} A_{ij}^2$$

$$= (D_1 D_2 D_3)^{-1/2} A_{12}^2 + (D_2 D_3 D_4)^{-1/2} A_{23}^2 = 1.000$$

To account for the nature of the atoms and the unsaturation of bonds in $\chi_R(G)$, *the valence molecular connectivity index* $\chi_R^v(G)$ was proposed.[3] The atom connectivity Δ_i^v is used instead of the valency of a vertex i, D_i, in calculating $\chi_R^v(G)$. The atom connectivity is defined as,

$$\Delta_i^v = Z_i^v - H_i \tag{9}$$

where Z_i^v represents the number of valence electrons of atom *i*, while H_i stands for the number of hydrogen atoms attached to it. Thus, the zero-, first-, and second-order valence molecular connectivity indexes may be calculated by means of the following formulae,

$$^0\chi_R^v(G) = \sum_{s=1}^{S_v} \left(\Delta_i^v\right)_s^{-1/2} \tag{10}$$

$$^1\chi_R^v(G) = \sum_{s=1}^{S_e} \left(\Delta_i^v \Delta_j^v\right)_s^{-1/2} ; i \neq j \tag{11}$$

Table 1

CONNECTIVITIES Δ_i^v FOR CARBON, NITROGEN, AND OXYGEN ATOMS

			Atom i			
Number of C-i bonds	Carbon	Δ_i^v	Nitrogen	Δ_i^v	Oxygen	Δ_i^v
0	CH_4	0	NH_4^+	1	OH_3^+	3
			NH_3	2	OH_2	4
1	$-CH_3$	1	$-NH_3^+$	2	$-OH_2^+$	4
			$-NH_2$	3	$-OH$	5
2	$=CH_2$	2	$=NH_2^+$	3	$=OH^+$	5
			$=NH$	4	$=0$	6
3	$\equiv CH$	3	$\equiv NH^+$	5	$\equiv 0^+$	7
			$\equiv N$	5		
4	$\equiv C$	4	$\equiv N^+$	6		

$$^2\chi_R^v(G) = \sum_{s=1}^{s_L} \left(\Delta_i^v \Delta_j^v \Delta_k^v\right)_s^{-1/2}; i \neq j \neq k \qquad (12)$$

Symbols in the above formulae have their previous meaning. For saturated hydrocarbons, formulae (10) to (12), because of identity $D_i = \Delta_i^v$, reduce to (5) to (7). Values of atom connectivities Δ_i^v for carbon, nitrogen, and oxygen in different chemical environments are given in Table 1. Note, that we will sometimes indicate the connectivity index only by the symbol χ.

3. The Maximal Eigenvalue Index

The maximal eigenvalue of the graph spectrum, x_1, may be used as a topological index,[6,9,10] because it is closely related to the number of walks, $w(L)$, in the graph[11] and for the regular graphs the following equation holds,

$$w(L) = N (x_1)^L \qquad (13)$$

This equation is derived in the following way. We start with the relation between the number of walks between the vertice p and q and the adjacency matrix,[11]

$$w_{pq}(L) = \left(A^L\right)_{pq} \qquad (14)$$

The total number of walks of the length L in the graph G is given by,

$$w(L) = \sum_{p=1}^{N} \sum_{q=1}^{N} w_{pq}(L) \qquad (15)$$

Using the relations,

$$C_j A = x_j C_j; \ j = 1,2,\ldots,N \qquad (16)$$

and

$$CA^L C^T = \text{diag}\left[(x_1)^L, (x_2)^L,\ldots,(x_N)^L\right] \qquad (17)$$

Equation (14) transforms to,

$$w_{pq}(L) = \sum_{j=1}^{N} (x_j)^L c_{jp} c_{jq} \qquad (18)$$

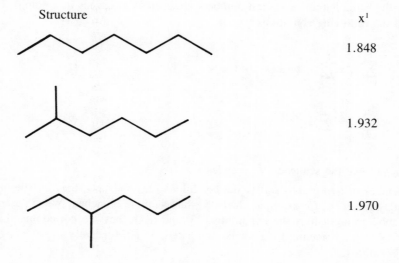

Introduction of (18) into (15) leads to the relation for the total number of walks in a more convenient form for our further discussion,

$$w(L) = \sum_{j=1}^{N} (x_j)^L \left(\sum_{p=1}^{N} c_{jp} \right)^2 \tag{19}$$

The eigenvectors of a regular graph have the following property,

$$\sum_{p=1}^{N} c_{jp} = \begin{cases} \sqrt{N} & \text{if } j=1 \\ 0 & \text{if } j>1 \end{cases} \tag{20}$$

Example

G

N = 3

$\sqrt{N} = 1.73205$

$C_{11} + C_{12} + C_{13} = 3 \cdot 0.57735 = 1.73205$

$C_{21} + C_{22} + C_{23} = -0.81650 + 0.40825 + 0.40825 = 0$

For regular graphs ($D_{min} = D_{max} = D$) we obtain from Frobenius theorem (see Chapter 5, Volume I, Section I),

$$D = x_1 \tag{21}$$

This result together with (20) leads to the simple expression for the total number of walks in a graph,

$$w(L) = D^L \cdot N = N \cdot (x_1)^L \tag{22}$$

For nonregular graphs this equation holds only approximately. Equation (22) shows that there is a close relationship between a spectral (x_1) and a combinatorial ($w[L]$) property of a graph. Thus, x_1 may be used to order chemical structures, albeit in some cases the same x_1 value will belong to two different systems.

Example

Structure	x^1
	1.848
	1.932
	1.970

Structure x^1

                                                        2.000

                                                        2.000

                                                        2.053

                                                        2.101

                                                        2.136

                                                        2.175

This example shows that the ordering of isomeric alkanes according to increasing value of x_1 follows the intuitive notion of branching.

4. The Comparability Index

The comparability index is proposed by Gutman and Randić.[12] These authors have used the algebraic concept of comparability of functions, introduced by Muirhead,[13] in order to compare acyclic graphs and to discuss the concept of branching. Slightly modifying their definition, we can express it as follows: Let $V = v_1, v_2, \ldots, v_N$ and $V' = v'_1 \ v'_2 \ \ldots, v'_N$ be two nonincreasing (i.e., $v_i \geq v_i + 1$ and $v'_i \geq v'_{i+1}$ for $i = 1, 2, \ldots, N - 1$) sequences of the same length of natural numbers (valencies of vertices in a graph). Let V and V' fulfill the following conditions,

$$\sum_{i=1}^{k} v_i \geq \sum_{i=1}^{k} v'_i \text{ for all k, } 1 \leq k \leq N \qquad (23)$$

$$\sum_{i=1}^{N} v_i = \sum_{i=1}^{N} v'_i \qquad (24)$$

Then we can say that the sequence V precedes V'.

When valencies of two graphs satisfy the conditions (23) and (24), they can be ordered and accordingly it can be, for example, decided which is more branched. When parameters v_i of two graphs do not satisfy the conditions (23) and (24), they are not comparable and their properties (e.g., branching) cannot be compared. When for every $i : v_i = v'_i$, graphs cannot be discriminated.

Example

1544

3064

A

$$\left\{\, 44411111111 \,\right\}$$

B

$$\left\{\, 44222111111 \,\right\}$$

C

$$\left\{\, 43331111111 \,\right\}$$

It follows that A is comparable to B and C, and can be ordered to precede them, but B and C are not comparable among themselves. The conclusion is that the structure A is more branched then either B or C. The same decision cannot be made for structures B and C.

Let us now compare graphs G and H.

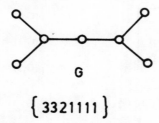

G

$$\left\{\, 3321111 \,\right\}$$

The comparability index {3321111} is not a unique representation of G, because structure H have the same symbols.

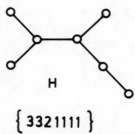

H

$$\left\{\, 3321111 \,\right\}$$

Thus, because for all $i:v_i = v_i'$, graphs G and H cannot be differentiated.

In summary, the analysis via the comparability index reveals that some graphs (structures) can be compared, some cannot, and some cannot be discriminated when the valency of the vertices involved is the sole basis for classification and ordering.

5. The Z Index

The Z index of a graph z(G) was introduced by Hosoya[14] and it is defined as,

$$Z(G) = \sum_{k=0}^{[N/2]} p(G;k) \qquad (25)$$

where the numbers $p(G; k)$ represent the number of ways in which k disconnected K_2 components can be imbedded into G as a subgraph. $[N/2]$ is the maximal *k* number for G. Note, $p(G; 0) = 1$ by definition, $p(G; 1) = $ number of edges, and $p(G; N/2) = $ number of 1-factors (Kekulé structures).

Example

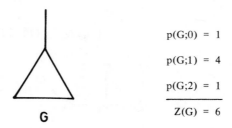

G

$$p(G;0) = 1$$
$$p(G;1) = 4$$
$$p(G;2) = 1$$
$$\overline{}$$
$$Z(G) = 6$$

The connection of the Z index and the adjacency matrix is through the Hosoya polynomial[14,15] or the Z-counting polynomial,

$$P^h(G;x) = \sum_{k=0}^{[N/2]} p(G;k) \, x^k \qquad (26)$$

The Z index may be obtained from the polynomial (26) for $x = 1$,

$$Z(G) = P^h(G;x=1) \qquad (27$$

Example

G

$$p(G;0) = 1$$
$$p(G;1) = 4$$
$$p(G;2) = 2$$

$$P^h(G;x) = 1 + 4x + 2x^2$$

$$P^h(G;x=1) = Z(G) = 7$$

The connection between the Z-counting polynomial and the characteristic polynomial is rather simple, thus indicating that it could be obtained from the adjacency matrix:

1.   For polyenes, because of identity, (16), is

$$P(L_N;x) = x^N P^h(L_N; -x^{-2}) \qquad (28)$$

2.   For annulenes it is given by,

$$P(C_N;x) = x^N P^h(C_N; -x^{-2}) - 2 \qquad (29)$$

3.   For polycyclic systems it is somewhat more complicated,

$$P(G;x) = x^N P^h(G;-x^{-2}) - 2x^{N-m} \sum_m P^h(G-C_m;-x^{-2}) +$$

$$4x^{N-(m+n)} \sum_{m<n} P^h(G-C_m-C_n;-x^{-2})\ldots \qquad (30)$$

Example

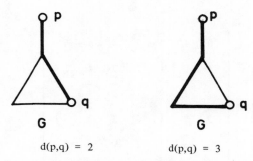

$$N = 10$$

G

$$A(G) \implies \det \left| xI - A(G) \right| \implies \sum_{n=0}^{N} a_n x^{N-n} \equiv P(G;x)$$

$$P(G;x) = x^{10} - 11x^8 + 41x^6 - 65x^4 + 43x^9 - 9$$

$$P(G;x) = x^{10}P^h(G;-x^{-2}) - 4x^4 + 12x^2 - 6$$

$$P^h(G;-x^{-2}) = 1 - 11x^{-2} + 41x^{-4} - 61x^{-6} + 31x^{-8} - 3x^{-10}$$

$$P^h(G-C_6;-x^{-2}) = 1 - 3x^{-2} + x^{-4}$$

$$\underline{P^h(G-C_{10};-x^{-2}) = 1}$$

$$P^h(G;-x^{-2}=x) = 1 + 11x + 41x^2 + 61x^3 + 31x^4 + 3x^5$$

$$P^h(G;x=1) \equiv Z(G) = 148$$

B. Indexes Based on Topological Distances

The distance matrix of the graph G, $D(G)$, contains information on distances in G. Early topological indexes, which were used for relating structure to properties of alkanes, were based on the distance matrix albeit their authors were not aware of this fact.[16,17] Several indexes based on *topological distances* will be described below following order of their appearance in literature. Let us remind the reader that the topological distance $d(p,q)$ is defined as the smallest number of edges (bonds) — shortest path — between the vertices (atoms) p and q in the graph (molecule).

Example

G	**G**
$d(p,q) = 2$	$d(p,q) = 3$

Therefore, $d(p,q) = 2$ is topological distance between vertices p and q in G.

1. The Wiener Number

Wiener,[16] in his studies on physical parameters of acyclic hydrocarbons, introduced a *path number* W. The path number W is defined as the number of bonds between all pairs of atoms in a acyclic molecule. In view of the pioneering contribution of Wiener in recognizing the significance of the number of paths in a molecular skeleton (i.e., acyclic skeletons) it seems appropriate to continue to call the number of distances in all (acyclic and cyclic) structures the *Wiener number* W(G). However, one should be aware that one-to-one correspondence between the pair of neighbors certain number of bonds away and the number of paths of the same length holds *only* for acyclic systems. Hence, in polycyclic structures Wiener number is associated with distance only, not with the number of paths, but in acyclic structures the two are the same.

The Wiener number of G is equal to the half-sum of the elements of **D**(G) (see Chapter 4, Volume I, Section IV).

$$W(G) = \frac{1}{2} \sum_{(i,j)} D_{ij}(G) \tag{31}$$

where $D_{ij}(G)$ represent off-diagonal elements of **D**(G). The smaller the Wiener number, the larger the compactness of the molecule.

Rouvray[18] proposed as a simple topological index the sum of all off-diagonal elements of the distance matrix,

$$R\,(G) = \sum_{(i,j)} D_{ij}(G) \tag{32}$$

Evidently,

$$R\,(G) = 2W(G) \tag{33}$$

Wiener[16] has also introduced the *polarity number* of an alkane, $p(G)$, which is equal to the number of pairs of vertices separated by three edges,

$$p(G) = \frac{1}{2} \Sigma\, W_3(G) \tag{34}$$

where $W_3(G)$ is the number of off-diagonal elements of **D**(G) with the distance 3. The polarity number indicates the number of steric pairs for a given alkane.

Example

$$\mathbf{D}(G) = \begin{array}{c} 1 \\ 2 \\ 3 \\ 4 \\ 5 \\ 6 \\ 7 \end{array} \begin{bmatrix} 0 & 1 & 2 & ③ & 4 & 5 & 6 \\ 1 & 0 & 1 & 2 & ③ & 4 & 5 \\ 2 & 1 & 0 & 1 & 2 & ③ & 4 \\ ③ & 2 & 1 & 0 & 1 & 2 & ③ \\ 4 & ③ & 2 & 1 & 0 & 1 & 2 \\ 5 & 4 & ③ & 2 & 1 & 0 & 1 \\ 6 & 5 & 4 & ③ & 2 & 1 & 0 \end{bmatrix}$$

$$W\,(G) = 8 \Longrightarrow p(G) = 4$$

2. The Platt Number

Platt[17] was also interested in devising a scheme for predicting the physical parameters (molar refractions, molar volumes, boiling points, heats of formation, heats of vaporization) of alkanes. He introduced an index F(G), which is equal to the total sum of degrees of edges in a graph. The degree of an edge e, $D(e)$, is the number of the adjacent edges. Therefore, the Platt number of G is defined by,

$$F(G) = \sum_{i=1}^{M} D(e_i) \qquad (35)$$

The Platt number, thus represents the first neighbors sum.

Example

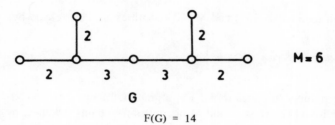

M = 6

$$F(G) = 14$$

3. The Gordon-Scantlebury Index

The Gordon-Scantlebury index[19] is defined as the number of distinct ways in which the chain fragment L_3, called *links*, can be embedded on the carbon skeleton of a given molecule,

$$S(G) = \sum_{i} (L_3)_i \qquad (36)$$

The Gordon-Scantlebury index thus represents the total number of 2-walks between different vertices of G,

Example

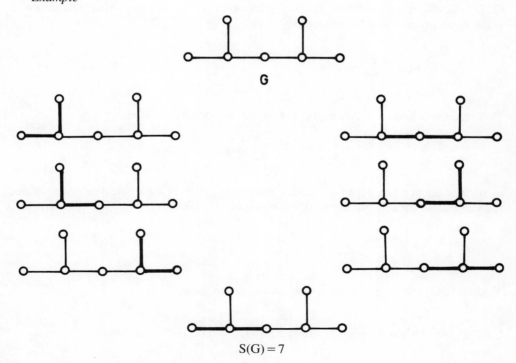

$$S(G) = 7$$

The Platt number and the Gordon-Scantlebury index are closely related quantities. It is easily shown that the $S(G)|$index is equal to the half the value of $F(G)$,

$$S(G) = \frac{1}{2} F(G) \tag{37}$$

Since the Gordon-Scantlebury index may also be expressed in terms of one vertex degree,

$$S(G) = \frac{1}{2} \sum_{j=1} D_j (D_j - 1) \tag{38}$$

it is not difficult to see that $S(G)$ and $M_1(G)$ quantities are also related by,

$$S(G) = \frac{1}{2} (M_1 - 2M) = \frac{1}{2} M_1 - (N - 1) \tag{39}$$

where M is the number of edges (bonds) in a graph (structure). D_j and M values of G are related by Equation (14), while N and M quantities for chains (alkanes) by $M = N - 1$.

4. The Altenburg Polynomial

Altenburg[20] proposed a following polynomial for characterization of a given structure,

$$P^A(G;x) = \sum_{j=1}^{N} a_j x_j \qquad a_j x^j \tag{40}$$

where a_j is the number of pairs of atoms separated by the distance j, while N is the maximal distance (diameter of G). Therefore, the coefficients of $P^A(G; x)$ measure the population of a given topological distance in G. Thus, they may be obtained from the distance matrix of G. In addition, if x_js are substituted by j-values ($j = 1, 2, \ldots N$), the Wiener number of a graph is simply obtained by multiplying a_j with j,

$$W(G) = \sum_{j=1}^{N} a_j \cdot j \tag{41}$$

This result was known to Altenburg, but he did not identify the sum

$$\sum_{j=1}^{N} a_j \cdot j$$

as the Wiener number. However, he has used these numbers for a calculating the squared radii of alkanes, $R^2(G)$, by means of the relation,

$$R^2(G) = \frac{1}{N^2} \sum_{j=1}^{N} a_j \cdot j \tag{42}$$

Equation (62) can be given, of course, in a different form,

$$R^2(G) = \frac{W(G)}{N^2} \tag{43}$$

Example

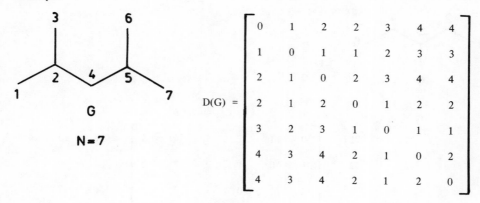

$$PA(G;x) = 6x_1 + 7x_2 + 4x_3 + 4x_4$$

$$W(G) = 6 \cdot 1 + 7 \cdot 2 + 4 \cdot 3 + 4 \cdot 4 = 48$$

$$R^2(G) = \frac{48}{49} = 0.9796$$

Altenburg has recently shown that there is a linear correlation between $R^2(G)$ and the Randić connectivity index for alkanes.[21]

5. The Balaban Index

Balaban[22] has recently proposed a topological index, named *distance sum connectivity index*, denoted by J(G), which appears to be the most discriminating single topological index so far proposed. Only the "superindex" of Bonchev et al.,[23] which is a sequence of several topological indexes, has comparative ability for differentiating chemical structures. The distance sum connectivity index of a graph is defined as follows,

$$J(G) = \frac{q}{\mu(G)+1} \sum_{edges} (d_i d_j)^{-1/2} \qquad (44)$$

where d_is ($i = 1,2,\ldots, N$) are the distance sums, q is the number of edges, while $\mu(G)$ is the cyclomatic number of a graph. The *distance sum* for a vertex i, d_i, represents a sum of all entries in the row or column of the distance matrix corresponding to that vertex,

$$d_i = \sum_{j=1}^{N} D_{ij} \qquad (45)$$

The *cyclomatic number* $\mu(G)$ of a graph indicates the number of its cycles and is equal to the minimal number of edges necessary to be removed in order to convert the (poly)cyclic graph to the related acyclic graph,

$$\mu(G) = q - N + 1 \qquad (46)$$

Example

$q = 5$

$\mu = 0$

$$\mathbf{D}(G) = \begin{array}{c} \\ 1 \\ 2 \\ 3 \\ 4 \\ 5 \\ 6 \end{array} \begin{array}{cccccc} 1 & 2 & 3 & 4 & 5 & 6 \\ \left[\begin{array}{cccccc} 0 & 1 & 2 & 3 & 4 & 3 \\ 1 & 0 & 1 & 2 & 3 & 2 \\ 2 & 1 & 0 & 1 & 2 & 1 \\ 3 & 2 & 1 & 0 & 1 & 2 \\ 4 & 3 & 2 & 1 & 0 & 3 \\ 3 & 2 & 1 & 2 & 3 & 0 \end{array}\right] \end{array} \begin{array}{l} d_1 = 13 \\ d_2 = 9 \\ d_3 = 7 \\ d_4 = 9 \\ d_5 = 13 \\ d_6 = 11 \end{array}$$

$$J(G) = 5\left[(13\cdot9)^{-1/2} + (9\cdot7)^{-1/2} + (7\cdot9)^{-1/2} + (7\cdot11)^{-1/2} + (9\cdot13)^{-1/2}\right] = 2.7542$$

$q = 6$

$\mu(G) = 2$

$$\mathbf{D}(G) = \begin{array}{c} \\ 1 \\ 2 \\ 3 \\ 4 \\ 5 \end{array} \begin{array}{ccccc} 1 & 2 & 3 & 4 & 5 \\ \left[\begin{array}{ccccc} 0 & 1 & 2 & 2 & 1 \\ 1 & 0 & 1 & 1 & 2 \\ 2 & 1 & 0 & 1 & 2 \\ 2 & 1 & 1 & 0 & 1 \\ 1 & 2 & 2 & 1 & 0 \end{array}\right] \end{array} \begin{array}{l} d_1 = 6 \\ d_2 = 5 \\ d_3 = 6 \\ d_4 = 5 \\ d_5 = 6 \end{array}$$

$$J(G) = \frac{6}{2+1}\left[(6\cdot5)^{-1/2} + (6\cdot6)^{-1/2} + (5\cdot6)^{-1/2} + (5\cdot5)^{-1/2} + (6\cdot5)^{-1/2} + (5\cdot6)^{-1/2}\right]$$

$$= 2.1939$$

In order to consider multigraphs (molecular structures with multiple bonds), Balaban suggested that the entries in the distance matrices of such a graph should be in terms of bond orders instead of topological distances. Thus, the entry of a single connection (a single bond) should be 1, the entry for a double connection (a double bond) 1/2, etc.

Example

G

$$q = 6$$

$$\mu(G) = 1$$

$$
D_b(G) = \begin{array}{c c}
 & \begin{array}{c c c c c c} 1 & 2 & 3 & 4 & 5 & 6 \end{array} \\
\begin{array}{c} 1 \\ 2 \\ 3 \\ 4 \\ 5 \\ 6 \end{array} &
\left[\begin{array}{c c c c c c}
0 & 1 & 1.5 & 2.5 & 3 & 2 \\
1 & 0 & 0.5 & 1.5 & 2 & 1 \\
1.5 & 0.5 & 0 & 1 & 2 & 1.5 \\
2.5 & 1.5 & 1 & 0 & 1 & 2 \\
3 & 2 & 2 & 1 & 0 & 1 \\
2 & 1 & 1.5 & 2 & 1 & 0
\end{array} \right]
\end{array}
\begin{array}{l}
d_1 = 10 \\
d_2 = 6 \\
d_3 = 6.5 \\
d_4 = 8 \\
d_5 = 9 \\
d_6 = 7.5
\end{array}
$$

$$
J(G) = 3 \left[(10 \cdot 6)^{-1/2} + (6 \cdot 6.5)^{-1/2} + (6 \cdot 7.5)^{-1/2} + (6.5 \cdot 8)^{-1/2} + (8 \cdot 9)^{-1/2} + \right.
$$

$$
\left. (9 \cdot 7.5)^{-1/2} \right] = 2.4495
$$

6. The Smolenskii Additivity Function

Smolenskii[24] has set forth a powerful procedure in order to study the additive properties of hydrocarbons. Smolenskii's procedure is based on the decomposition of a hydrogen-suppressed graph into fragments of various types $|X_k^{P_k}|$. The fragments of a given type then become a variable in a multiple regression equation. A linear function of degree N is defined for graph G,

$$
f(G) = a_0 + \sum_{k=1}^{N} \sum_{P=1}^{S_N} a_k^{P_k} \left| X_k^{P_k} \right| \tag{47}
$$

The coefficients $a_K^{P_k}$ are determined by multiple regression against the experimental values for a given molecular property. The quantity S_N gives the maximum count for a given type of fragments of k edges associated with P_k edges of the graph. The count of each set of fragments is given by the quantity $|X_K^{P_k}|$. The list of some of the Smolenskii fragments is given in Figure 1.

An example of the enumeration of the Smolenskii fragments is shown below for 2,2,4,4-tetramethylpentane.

Symbol   **Fragment**

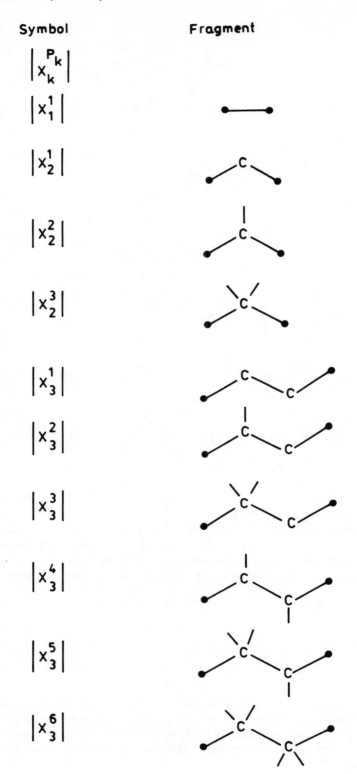

FIGURE 1.   The Smolenskii fragments.

Example

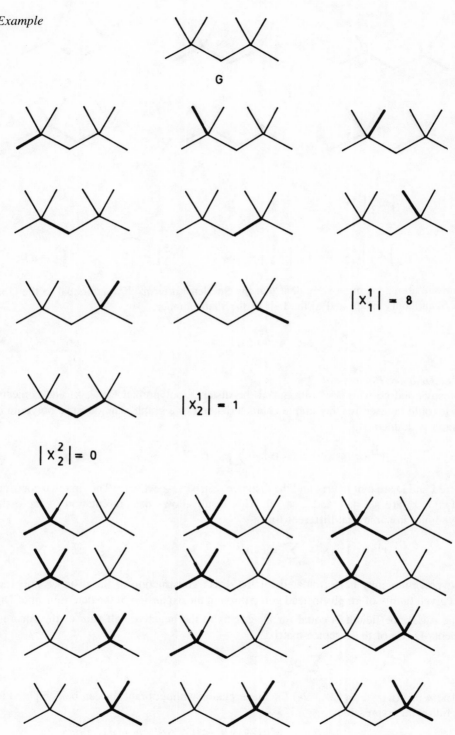

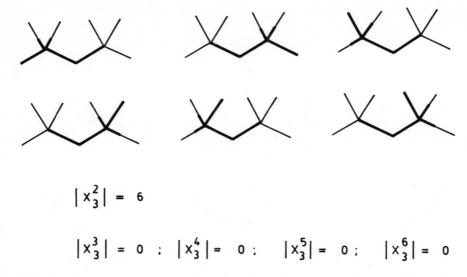

$$\left| x_3^2 \right| = 6$$

$$\left| x_3^3 \right| = 0 \; ; \quad \left| x_3^4 \right| = 0 \; ; \quad \left| x_3^5 \right| = 0 \; ; \quad \left| x_3^6 \right| = 0$$

From the above example is evident that the Smolenskii terms $|X_2^{P_K}|$ are equal to the Gordon and Scantlebury index and to the half of the Platt number,

$$\sum_{p=1}^{S_N} \left| X_2^{P_k} \right| = S(G) = \frac{1}{2} F(G) \tag{48}$$

7. The Distance Polynomial

Hosoya and co-workers[25] introduced the distance polynomial $P^D(G; x)$ and conjectured that it could be used for the unique characterization of a graph. The distance polynomial of a graph is defined as,

$$P^D(G;x) = \det \left| xI - D \right| = \sum_{n=0}^{N} a_n(G)x^{N-n} \tag{49}$$

where I and D are unit matrix and the distance matrix, respectively. The first two coefficients of $P^D(G; x)$ are $a_0 = 1$ and $a_1 = 0$. This result allows the presentation of the distance polynomial in somewhat different form,

$$P^D(G;x) = x^N - \sum_{n=2}^{N} a_n(G)x^{N-n} \tag{50}$$

Several rules may be formulated for the direct construction of the distance polynomial for a given family of graphs instead going through the expansion of the determinant of $D(G)$.

The $a_2(G)$ coefficient is equal for all graphs to the negative half-sum of the squares of elements $D_{ij}(G)$ of the distance matrix,

$$a_2(G) = -\frac{1}{2} \sum_{(i,j)} D_{ij}^2 \tag{51}$$

The $a_n(G)$ (n = 2,3,4,. . .,N) for linear chains without branches can be computed using the following expression,

$$a_n(G) = 2^{n-2}(n-1)\frac{N^2(N^2-1)(N^2-2^2)\dots[N^2-(n-1)^2]}{n^2(n^2-1)(n^2-2^2)\dots[n^2-(n-1)^2]} \tag{52}$$

which for n=2 reduces to,

$$a_2(G) = \frac{N^2(N^2-1)}{4\cdot3} \tag{53}$$

This equation produces, of course, identical values of $a_2(G)$ for acyclics, as Equation (51).

Expressions for $a_n(G)$ coefficients of other classes of graphs are very complicated. However, the explicit expressions for the last two coefficients of the distance polynomial, $a_{N-1}(G)$ and $a_N(G)$, are available for several families of graphs.

(a) Trees

$$a_{N-1} = 2^{N-3} \sum_D v_D(D-1)(2N-D+2) \tag{54}$$

where v_D stands for the number of vertices with D degrees,

$$a_N = 2^{N-2}(N-1) \tag{55}$$

The above relation can be directly obtained from (52) for n = N.

Example

G **N = 5**

$$a_0(G) = 1$$

$$a_2(G) = \frac{5^2(5^2-1)}{4\cdot 3} = 50$$

$$a_3(G) = 2(3-1)\frac{5^2(5^2-1)(5^2-2^2)}{3^2(3^2-1)(3^2-2^2)} = 140$$

$$a_4(G) = 2^2\ [3\cdot 1\cdot(10-2+2)] = 120$$

$$a_5(G) = 2^{5-2}(5-1) = 32$$

$$p^D(G;x) = x^5 - 50x^3 - 140x^2 - 120x - 32$$

(b) Cycles

For cycles the expression for $a_{N-1}(G)$ is rather unwieldy, while for $a_N(G)$ depends on the ring size.

$$a_N(G) = \begin{cases} 2^{N-4}N^2 & N = \text{even} \\ (N^2-1)/4 & N = \text{odd} \end{cases} \tag{56}$$

(c) Cycles with branches

Here again the expression for $a_{N-1}(G)$ is very complicated, whereas for $a_N(G)$ depends again on the ring size.

$$a_N(G) = \begin{cases} 2^{N-3}(N\cdot m - m^2/2) & m = \text{even} \\ 2^{N-m-1}[N\cdot m - (m^2+1)/2] & m = \text{odd} \end{cases} \tag{57}$$

where m denotes an m-membered cycle.

8. *The Information Indexes*

Bonchev and Trinajstić[26] applied information theory to the problem of characterizing molecular structures.[27-30] Information theory provides a simple quantitative measure called the *information content* of a given system. The information content of a system, I, with N elements is defined by the relation,[31]

$$I = N\log_2 N - \sum_{j=1}^n N_j \log_2 N_j \tag{58}$$

where n is the number of different sets of elements, N_j is the number of elements in the j-th set of elements, and summation is over all sets of elements. The logarithm is taken at basis 2 for measuring the information content in bits. Another information measure is the *mean information content* of one element of the system, I, defined by means of the total information content or by the Shannon relation,[32]

$$\bar{I} = \frac{I}{N} = -\sum_{j=1}^{n} P_j \log_2 P_j \qquad (59)$$

where $P_j = N_j/N$.

The application of information theory to different systems or structures is based on the possibility of constructing a finite probability scheme for every system. For example, the distance matrix of a given graph $D(G)$ has a convenient structure of the application of information theory. The elements of distance matrix $D_{ij}(G)$ can be considered as the element of a finite probability scheme associated with G. A certain distance of a value j ($1 \leqslant j \leqslant N - 1$) appears $2a_j$ times in the distance matrix. All elements of $D(G)$ may be partitioned into $n + 1$ groups. The $n + 1$ group would contain only diagonal elements of $D(G)$ which are, of course, all equal zero. Since the distance matrix is symmetric matrix, the upper-triangle part of its preserves all the information about the corresponding system. The total number of elements in the upper triangle of $D(G)$ is $N(N-1)/2$.

Statistical analysis of distance matrix on the basis of information theory leads to the following topological indexes: (1) the information content and mean information content on realized distances in the graph G, $I_D^W(G)$ and $\bar{I}_D^W(G)$, and (2) the information content and mean information content on the distribution on the distances in G, $I_D^E(G)$ and $\bar{I}_D^E(G)$. They are defined as follows,[26,33]

$$I_D^W(G) = W \log_2 W - \sum_{j=1}^{n} a_j \cdot j \log_2 j \qquad (60)$$

$$\bar{I}_D^W(G) = \frac{I_D^W(G)}{W} = -\sum_{j=1}^{n} a_j \frac{j}{W} \log_2 \frac{j}{W} \qquad (61)$$

$$I_D^E(G) = \frac{N(N-1)}{2} \log_2 \frac{N(N-1)}{2} - \sum_{j=1}^{n} a_j \log_2 a_j \qquad (62)$$

$$\bar{I}_D^E(G) = \frac{2I_D^E(G)}{N(N-1)} = -\sum_{j=1}^{n} \frac{2a_j}{N(N-1)} \log_2 \frac{2a_j}{N(N-1)} \qquad (63)$$

Example

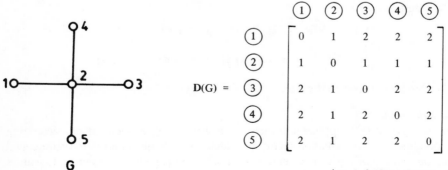

$$D(G) = \begin{array}{c} \\ \textcircled{1} \\ \textcircled{2} \\ \textcircled{3} \\ \textcircled{4} \\ \textcircled{5} \end{array} \begin{bmatrix} 0 & 1 & 2 & 2 & 2 \\ 1 & 0 & 1 & 1 & 1 \\ 2 & 1 & 0 & 2 & 2 \\ 2 & 1 & 2 & 0 & 2 \\ 2 & 1 & 2 & 2 & 0 \end{bmatrix}$$

j : 1, 2 (Only the upper triangle of the distance matrix is considered)

a_1 : 4

a_2 : 6

$N = 5, \quad \frac{N(N-1)}{2} = 10$

$$W = \sum_{j=1} a_j \cdot j = 4 \cdot 1 + 6 \cdot 2 = 16$$

$$I_D^W(G) = 16 \log_2 16 - 6 \cdot 2 \log_2 2 = 52 \text{ bits}$$

$$\bar{I}_D^W(G) = \frac{I^W(G)}{W} = \frac{52.00}{16} = 3.25 \text{ bits}$$

$$I_D^E(G) = 10 \log_2 10 - \left\{ 4 \log_2 4 + 6 \log_2 6 \right\} = 9.71 \text{ bits}$$

$$\bar{I}_D^E(G) = \frac{2 I_D^E(G)}{N(N-1)} = \frac{9.71}{10} = 0.97 \text{ bits}$$

9. The Centric Index

In order to have a topological index which reflects the *shape* of an alkane, Balaban[34] has introduced the *centric index*. Let us denote for a given vertex v_j,

$$r_j = \max_{v_i \in V(G)} d(v_j, v_i) \tag{64a}$$

where r_j is the maximum possible distance between v_j and any other vertex in G. Then,

$$r = \min_{v_j \in V(G)} r_j \tag{64b}$$

is called the *radius of the graph* and each vertex v_j such that $r_j = r$ is called a central point. The *center* of G is the set of all central points. A classical result of graph theory is that any tree has a center consisting of either one point of two adjacent points.[35]

Normalized centric index $C(G)$ is given by,

$$C(G) = \frac{1}{2} [B(G) - 2N + U] \tag{65}$$

where

$$B(G) = \sum_{j=1} \delta_i^2 \tag{66}$$

$$U = [1 - (-1)^N]/2 \tag{67}$$

Numbers δ_i represent a pruning sequence. Under the pruning of a tree graph, the stepwise process of removing all vertices of degree one (end points) from G together with their incident edges until the center of G is found is understood. The number of vertices removed at each step constitute the pruning sequence of a tree.

The centric index of a tree may be also defined as below,

$$C(G) = \frac{1}{2} [B(G) - B(\text{chain})] \tag{68}$$

where $B(\text{chain})$ gives B-quantity for a chain (*n*-alkane) with the same number of vertices as a branched tree (branched alkane).

Example

N = 7

G

1.   Pruning process

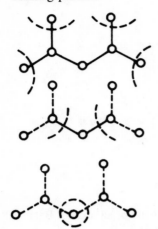

Number of endpoints
removed at each
pruning step

4

2

1

2.   Center of a graph

3.   Pruning sequence

(4,2,1)

4.   Centric index

$$B(G) = 4^2 + 2^2 + 1^2 = 21$$

$$U = [1 - (-1)^7]/2 = 1$$

$$C(G) = \frac{1}{2}(21 - 13) = 4$$

5.   Calculation of the centric index by means of Equation (68)

●B(G) = 21

● 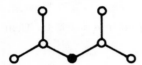   Chain with N = 7

●Pruning process gives the following pruning sequence:
(2,2,2,1)

●B(chain) = 13

●C(G) = $\frac{1}{2}$ (21 − 13) = 4

This example shows that the Balaban original procedure[34] for determining the center of a tree is applicable only to acyclic systems. For polycyclic graphs, no simple procedure for determining topological center could be found. However, recently the graph center concept was generalized and proposals for defining polycenter of any graph (acyclic or cyclic) are offered.[36,37] These proposals are based on the distance matrix of a graph.

The polycenter is the set defined by the following procedure:

1. Find the center of a graph
2. From the center, pick up points with the minimum sum of all distances from other vertices in G, i.e.,

$$\sum_{i=1}^{N} D_{ij} = \text{min.}$$

3. From these vertices, pick up points with the smallest number of times the maximum distance occurs and the denote the resulting set as C (pseudocenter).
4. Create a subgraph G' of G on the set C (V[G']=C) and repeat the procedure (1) to (3). This proceeds until in the *n*-th step, the same C as one at the beginning of this step is obtained. The set of points thus obtained is called the *polycenter* of G.

Determination of the polycenter by the above procedure is shown for several examples in Table 2. Note, that the polycenter of a graph may contain one vertex or several adjacent or nonadjacent vertices.

II. APPLICATIONS OF TOPOLOGICAL INDEXES

One of the ultimate targets of theoretical chemists is to build schemes that would allow accurate predictions of the bulk properties of matter from the knowledge of molecular structure. We are still far away from this ideal, but one way of trying to achieve this goal is by means of topological indexes since they serve as convenient descriptors of molecular structure. In this section, we will review some efforts in using topological indexes in the structure-property and structure-activity relationships. The standard strategy in devising QSPR or QSAR schemes with topological indexes consists of two steps:

1. Representation of a molecular structure by a given topological index
2. Finding the quantitative relationship between the topological indexes and physico-chemical properties (QSPR) or biological activities (QSAR) of the series of compounds under the investigation

A. Topological Indexes in the Quantitative Structure-Property Relationships

Among the first attempts to correlate physico-chemical properties with the structure of compounds is endeavored by Wiener.[16] Wiener found that the boiling points of alkanes may be correlated with their structures by means of the equation,

$$\Delta bp = (98/N^2)\Delta W + 5.5 \, \Delta p \tag{69}$$

where

$$\Delta bp = (bp)_0 - (bp) \tag{70}$$

$$\Delta W = W - W_0 \tag{71}$$

$$\Delta p = p_0 - p \tag{72}$$

$(bp)_0$ is the boiling point of the linear member of the group of isomers with the Wiener number W_0 and the polarity number p_0. For normal alkanes these structural variables are given by,

$$W_0 = \frac{N}{6} (N^2 - 1) \tag{73}$$

$$p_0 = N - 3 \tag{74}$$

Table 2
DETERMINATION OF CENTERS OF (POLY)CYCLIC GRAPHS ACCORDING TO THE PROCEDURE BY BONCHEV ET AL.[36,37]

Graph 1 (G)

	1	2	3	4	5	6	7	8	(i)	(ii)	(iii)	(iv)
1	0	1	2	3	2	1	2		3			
2	1	0	1	2	3	2	1		3			
3	2	1	0	1	2	3	2		3			
4	3	2	1	0	1	2	1		3			
5	2	3	2	1	0	1	2		3			
6	1	2	3	2	1	0	1		3			
7	2	1	2	1	2	1	0		2←			

Graph 2 (G)

	1	2	3	4	5	6	7	(i)	(ii)	(iii)	(iv)
1	0	1	2	3	1	2	2	3			
2	1	0	1	2	2	1	1	2	8←		
3	2	1	0	1	2	2	2	2	10		
4	3	2	1	0	1	1	3	3			
5	1	2	2	1	0	1	3	3			
6	2	1	2	1	1	0	2	2	9		
7	2	1	2	3	3	2	0	3			

Graph 3 (G)

	1	2	3	4	5	6	7	8	(i)	(ii)	(iii)	(iv)
1	0	1	2	2	3	4	5	4	5			
2	1	0	1	1	2	3	4	3	4			
3	2	1	0	2	3	4	5	4	5			
4	2	1	2	0	1	2	3	2	3	13	$1^2\,2^4\,3^1$←	
5	3	2	3	1	0	1	2	1	3	13	$1^3\,2^2\,3^2$	
6	4	3	4	2	1	0	1	2	4			
7	5	4	5	3	2	1	0	1	5			
8	4	3	4	2	1	2	1	0	4			

Graph 4 (G)

	1	2	3	4	5	6	(i)	(ii)	(iii)	(iv)
1	0	1	2	2	1	2	2	8		Repeat
2	1	0	1	2	2	1	2	7	$1^3\,2^2$	criteria
3	2	1	0	1	2	2	2	8		(i) to (iii)
4	2	2	1	0	1	1	2	7	$1^3\,2^2$	on psedo-
5	1	2	2	1	0	1	2	7	$1^3\,2^2$	graph G′
6	2	1	2	1	1	0	2	7	$1^3\,2^2$	

Graph G′ (nodes 1(2), 2(6), (5)4, 3(4))

	1	2	3	4	(i)
1	0	1	2	2	2
2	1	0	1	1	1←
3	2	1	0	1	2
4	2	1	1	0	2

*   Black points denote polycenters, whereas white points pseudocenters.

Example

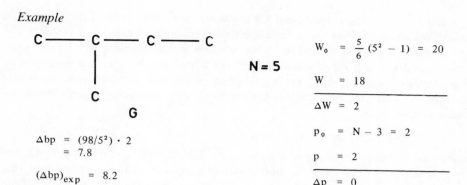

$$W_o = \frac{5}{6}(5^2 - 1) = 20$$

$$W = 18$$
_____
$$\Delta W = 2$$

$$P_o = N - 3 = 2$$

$$p = 2$$
_____
$$\Delta p = 0$$

$$\Delta bp = (98/5^2) \cdot 2$$
$$= 7.8$$

$$(\Delta bp)_{exp} = 8.2$$

Wiener[16] later developed a general expression for correlating properties of alkanes (molar refractions, molar volumes, boiling points at normal pressure, and heats of formation) denoted by P with the path number and polarity number,

$$P = \frac{a}{W} + b \cdot p \qquad (75)$$

where a and b are empirical adjustable parameters depending on the property P.

The Wiener number has also been used for predicting the chromatographic retention indexes of alkanes[38] and mono-alkyl- and di-alkylbenzenes.[39] The following two-parameter equation for calculating the retention indices of alkylbenzenes is obtained,

$$RI = a[W(G)]^n \qquad (76)$$

where parameters $a(244 \pm 4)$ and $n(0.297 \pm 0.03)$ were obtained by the least-squares fit for the sample of 28 molecules (the correlation coefficient $r = 0.9991$). Randić's approach is also applied to alkylbenzenes producing the linear least-squares relation (with correlation coefficient $r = 0.993$),[40]

$$RI(Rand\acute{i}c) = 132.8978 + 183.3725\ ^1\chi_R(G) \qquad (77)$$

Therefore, both topological indexes, the connectivity index and the Wiener number, appears to be convenient theoretical devices for calculating the chromatographic retention values from the structural parameters of molecules.

In addition, the reliability of the Wiener number for characterizing molecular branching[26,27] and cyclity[28,29] has been extensively studied. Likewise, the π-electron energy characteristics of conjugated infinite polymers were related to the Wiener number.[41-43]

Platt[17] tried to derive a predictive scheme, using a combination of the F(G) index with the Wiener number and/or the polarity number, for physico-chemical parameters such as molar refractions, molar volumes, boiling points, heats of vaporization, and heats of formation of alkanes. For example, Platt offered the following relation for predicting the molar refractions, R,

$$R = 6.460 + 4.750(N-1) + 0.0112F(G) - 0.134\ p \pm 0.16 \qquad (78)$$

The Rouvray index,[18] R(G) (a double Wiener number of a graph), was applied to the bulk thermodynamic properties of alkanes and alkenes (viscosity, surface tension, refractive index, boiling point, density), alkynes (boiling point, melting point, density and refractive index), and alkenes (boiling point). For correlating the thermodynamic property P and the R(G) index, a relationship of the general form is used,

$$P = a[R(G)]^b \qquad (79)$$

where a and b are constants determined for a given series. In all studied cases, the good agreement was obtained between the theoretical and experimental values.

The connectivity index of Randić has been extensively used in chemical[7,44-51] and pharmaceutical and biological calculations.[1,3,4,52-55] The correlation between the enthalpies of formation of alkane isomers, ΔH_f^o, and $^1\chi_R(G)$ gives, within each group of isomers, straight parallel branches emerging from the straight line correlating data of n-alkanes. This linear expression is given by,[7]

$$\Delta H_f^o = c_1 \, ^1\chi_R(G) + c_2 \cdot n + c_3 \qquad (80)$$

where the values for the above constants are: $c_1 = 12$ kcal/mol; $c_2 = -11$ kcal/mol; and $c_3 = -9$ kcal/mol. Several other thermodynamic properties of alkanes correlate well with $^1\chi_R(G)$:x7 boiling points, free energies, heats of solutions, parameters A, B, and C of the Antoine equation relating vapor pressure to temperatures, theoretically calculated total surface area of saturated acyclic hydrocarbons based on the model of Hermann,[56] etc. In addition, $^1\chi_R(G)$ correlates with Kováts empirical branching index,[57] derived from chromatographic retention data of isomeric alkanes.[7] Randić[44] has produced the following expression for calculating the retention indexes RI, of alkanes,

$$\text{RI(Randić)} = 200[^1\chi_R(G) - 1.4142 + (T_3)^2 + 300] \qquad (81)$$

where the factor 200 converts the Kováts retention scale to that of the connectivity values, while 1.4142 and 300 are the connectivity index and the Kováts retention value, for propane, respectively. T_3 stands for the number of terminal paths of length three, i.e., the number of paths between two methyl groups three bonds apart. Theoretical results obtained by the Randić relation (81) are in good agreement with the experimental values for alkanes. The Randić procedure for calculating the rentention indexes by means of $\chi_R(G)$ has been extended to several classes of compounds.[45-48,51,58,59]

In their series of papers on applications of molecular connectivity, Hall, Kier, Murray, Randić and others, found a significant correlation between the connectivity indexes and several physico-chemical properties such as, molar refraction,[8] diamagnetic susceptibility,[7] heat of vaporization,[7] boiling points,[61] liquid density,[62] water solubility,[61] and the partition coefficient for n-octanol/water.[63,64] Compounds used in these regression analyses were alkanes, alkenes, aliphatic alcohols, ethers, ketones, acids, esters, halides, and alkyl benzenes. The results of these correlations are summarized in Table 3.

Inclusion of the valence connectivity indexes into regression analysis resulted in appreciably improved correlation equations for heteroatomic compounds.

Finally, it has been found that the Hückel energy E_π of conjugated hydrocarbons parallels the valence molecular connectivity index.[65,66] The regression analysis produced linear correlations between E_π of alternant and nonalternant hydrocarbons and $^1\chi_R^v(G)$,

$$E_\pi(\text{alternants}) = 4.070 \, ^1\chi_R^v(G) - 0.302 \qquad (82)$$

with a correlation factor r = 0.999, and

$$E_\pi(\text{nonalternants}) = 4.068 \, ^1\chi_R^v(G) - 0.551 \qquad (83)$$

with a correlation factor r = 0.999.

Hosoya and co-workers[67] correlated the Z index with boiling points, bp, of saturated acyclic hydrocarbons. They found the following relation:

$$\text{bp} = [a \log Z(G) + b]/\sqrt{N} + c \qquad (84)$$

where the values of constants a, b, and c are dependent on the class of compounds under investigation. For example the relation of the boiling points (in °C) of alkanes at normal pressure is given by,

Table 3
MOLECULAR CONNECTIVITY AND PHYSICO-CHEMICAL PROPERTIES

Property	Class of compounds	Indexes	n^a	r^b	s^c
Molar refraction	Alcohols, ethers, amines, halides	$^1\chi; \, ^1\chi^V$	64	0.9897	1.01
	Alkyl-benzenes	$^1\chi; \, ^1\chi^V; \, ^2\chi; \, ^2\chi^V$	70	0.9985	0.163
Diamagnetic susceptibility	Alkanes	$^0\chi; \, ^1\chi; \, ^2\chi$	27	0.9998	0.415
Heat of vaporization	Alcohols	$^1\chi; \, ^1\chi^V$	20	0.9927	0.371
Boiling point	Alkanes	$^1\chi; \, ^4\chi_{PC}{}^d$	51	0.9851	5.51
				(0.9969)	(2.53)
	Alcohols	$^1\chi$	62	0.958	9.66
Liquid density	Alkanes	$1/^1\chi; \, ^3\chi_P{}^d$	82	0.9889	0.0046
	Alcohols	$^3\chi_P{}^d$	40	0.9246	0.0040
Water solubility	Alcohols	$^1\chi$	51	0.978	0.455
	Ethers	$^1\chi$	22	0.9895	0.242
	Esters	$^1\chi; \, ^1\chi^V$	38	0.9853	0.337
Partition coefficient	Alcohols, ethers, ketones acids, esters, amines	$^1\chi^V$	138	0.986	0.152
	Hydrocarbons	$^1\chi^V$	45	0.975	0.160

[a] Number of compounds.
[b] Correlation coefficient.
[c] Standard error.
[d] PC-path/cluster, P-path.

$$\text{bp(alkanes)} = [650.3 \log Z(G)]/\sqrt{N} + 46.55 \qquad (85)$$

This equation was compared with several empirical formulae available in the literature[68-71]

$$\text{bp} = 745.42 \log (N + 4.4) - 416.31 \qquad (86)$$

$$\text{bp} = \left\{ (21720N - 11820)/(1 + 0.000283N^2) \right\}^{1/2} \qquad (87)$$

$$\text{bp} = 127.55N^{2/3}/(1 + 0.07033N^{2/3}) \qquad (88)$$

$$\log (1078 - \text{bp}) = 3.0319 - 0.04999N^{2/3} \qquad (89)$$

The best correlation was obtained with Equation (89).

The Z index was also used for coding structures in computer-oriented chemical documentation.[72] In addition, Z indexes of many acyclic and cyclic structures are available in the form of tables.[73,74]

Smolenskii[24] has correlated the additivity function with standard heats of formation, ΔH_f^0, for 44 alkanes. The best regression relation is given by,

$$\Delta H_f^0 = 25.097 \left| X_2^1 \right| - 10.775 \left| X_2^2 \right| - 6.617 \left| X_2^3 \right| - 20.158 \left| X_3^1 \right| -$$

$$10.377 \left| X_3^2 \right| - 7.038 \left| X_3^3 \right| - 5.601 \left| X_3^4 \right| - 3.950 \left| X_3^5 \right| -$$

$$2.824 \left| X_3^6 \right| \qquad (90)$$

with a correlation coefficient r = 0.9984 and a standard error $\epsilon = 0.610$.

Nonlinear correlation between the information index $I_D^W(G)$, the number of carbon atoms N, and physical properties (heats of formation, heats of combustion — both in the liquid and gas phase — and vaporization, molecular volume, boiling point, parachor) P of linear and branched polyalkylbenzenes produced excellent results.[75] The correlations are of the general form,

$$P = a + b \cdot N + c \cdot I_{D,1}^{W}(G) + d \cdot N \cdot I_{D,1}^{W}(G) \tag{91}$$

where $I_{D,1}^{W}(G)$ is the information index $I_{D}^{W}(G)$ normalized to *one* topological distance,

$$I_{D,1}^{W}(G) = \frac{2I_{D}^{W}(G)}{N(N-1)} \tag{92}$$

In all cases studied, the correlation coefficient was found to be equal or close to unity, while the standard errors were 0.04% for heats of formation, 0.07 and 0.08% for heats of combustion in the liquid and gas phase, respectively, 0.4% for molecular volumes and heats of vaporization, 1.3% for boiling points, and 0.2% for parachor.

The normalized centric index gives a good correlation with octane numbers O_N of alkanes,[76]

$$O_N = a + b \cdot A_i(G) + c \cdot C(G); \quad i = 1,2 \tag{93}$$

where *a, b,* and *c* are statistical parameters while $A_i(G)$, $i = 1,2$, stands for the size of alkane expressed either as its number, N, of carbon atoms,

$$A_1(G) = N \tag{94}$$

or its molecular weight

$$A_2(G) = 14 \cdot N + 2 \tag{95}$$

Equation (93) may be presented in a more general form[76] to include any topological index, $T_j(G)$, proposed in literature,

$$O_N = a + b \cdot A_i(G) + c \cdot T_j(G); \quad i = 1,2$$

$$j = 1,2 \tag{96}$$

Balaban and Motoc[76] have used 14 topological indexes in the above relation obtaining the following results: (1) the centric index produced the best correlation with the octane number and (2) the intercorrelations between the various topological indexes indicated that the most of them are strongly intercorrelated.

The Randić comparability (similarity) concept[12,77,78] has been used in seeking similarities between the skeletal forms and connectivities in acyclic hydrocarbons. The method is illustrated on the trends of their thermodynamic properties such as boiling points and heats of vaporization.[79-82] This kind of the approach can be used to trace occasional deviations from the trends in the structure-property and structure-activity correlations.

B. Topological Indexes in the Quantitative Structure-Activity Relationship

The idea that there is a connection between the structure of a molecule and its biological action is rather old.[83] Since then QSAR have been elaborated in many different ways.[84-87] Thus, various mathematical models (Free-Wilson model,[88] Hanch model,[89] Lien model,[90] etc.) and many quantum-mechanical approaches have been proposed.[91-93] QSAR studies are certainly a major factor in contemporary drug design[84,85] and therefore represent an important field of research.

Since the size and shape of drug molecules are important factors for their biological response[3] and since topological indexes contain the information about the molecular size and shape, the application of topological indexes to QSAR appears to be a natural step in the development of the method. Therefore, this section will be devoted to correlations between biological properties and topological indexes. From the topological indexes, only the connectivity indexes have so far been extensively tested in QSAR studies, either alone or in different combinations with other physico-chemical or theoretical parameters. The use of connectivity indexes in QSAR studies was a natural consequence of their successful application in correlating the partition coefficients[63] of organic compounds between octanol and water with the calculated molecular polarizability.[8] It had already been found by Hansch and Fujita[94] that octanol/water partition coefficients of organic compounds well-reflected their ability to be transported through biological systems, which is responsible for a great deal of their biological activity. Molecular polarizabilities were also found to correlate with biological responses.[95,96] Thus, it was understandable to expect that the connectivity indexes would be appropriate descriptors for QSAR studies.

Kier and Hall[3,97] have considered a number of biological studies, using molecular connectivity indexes, in their effort to establish whether quantitative structure-activity relationships can be deduced by this new approach. They have demonstrated that structural characteristics, described by the molecular connectivity indexes, can be used to explain various nonspecific biological activities. By nonspecific activity, it is meant that a drug-receptor interaction is not regulated by strict structural requirements. Active molecules usually possess diverse structures and interact with the receptor over most of its surface rather than at selected functional groups arranged in a highly specific way (frequently called pharmacophores).

Anesthetic gases are quite varied in their structure and are considered as nonspecific agents. The first- and second-order valence molecular connectivity indexes have been calculated for a group of 28 gases.[98] This group consists of saturated hydrocarbons, perhalogenated and partially halogenated hydrocarbons, diethylether, SF_6, N_2O and $Cl_2CHCF_2OCH_3$. The linear one-variable ($^0\chi^v$) equation plotted for the anesthetic gases accounted for 78% of the variance in anesthetic pressure and the correlation coefficient was 0.881. The addition of a second variable (the electronic charge on the polar hydrogens), to describe the possibility of hydrogen bond formation, considerably improved the correlation (r = 0.966). Recently an additional study[99] has been made on a structurally similar group of general anesthetics (45 halogenated methane, ethane, propane and butane derivatives). Similarly to the previous study, the polar hydrogens were found to influence significantly the potency of anesthetic gases. The polarity of hydrogens was computed as a derivative of the Swain-Lupton **F** values[100] (the relative field effect of halogen atoms on the polarity of the C–H bond). A very good correlation was obtained (r = 0.975) between anesthetic activity and the valence zero-order molecular connectivity index ($^0\chi^v$) in addition to the polar hydrogen factor (the sum of all C–H bonds polarity terms). The predictive quality of the correlation equation obtained in this study has been illustrated by its ability to predict activities of 19 compounds not included in the initial investigation. Another group of general anesthetics whose structure-activity relationships have been examined by means of the molecular connectivity are anesthetic ethers[101] and a mixed group[102] of aliphatic hydrocarbons, ethers, and ketones. Statistical analysis showed that the anesthetic potency of 28 ethers[101] has an equally good parabolic (r = 0.986) and hyperbolic (r = 0.979) relationship with the connectivity term ($^1\chi$). The correlations have been obtained for a limited group of compounds, which has only two representatives past the parabola maximum and still very close to it. This explains why both correlation equations are equally good in describing the variation in anaesthetic activity. The toxicity (LD_{50}) of these ethers has also been examined by the molecular connectivity method and it was found that the number of edges (bonds) is the best single variable to correlate with it (r = 0.952). A very good correlation has been obtained[3,102] between anesthetic (AD_{100}) and toxic (LD_{100}) activities and the molecular connectivity index of a mixed class of general anaesthetics (27 aliphatic hydrocarbons, ethers, and ketones). In order to obtain good correlations for such a structurally diverse group of molecules, multiple regressions were necessary with highly complex connectivity indexes of the third-, fourth-, and sixth-order.

The study of local anesthetics[8] also illustrates the possibility of using molecular connectivity to describe nonspecific drug actions. The minimum blocking concentration of 36 various compounds which describes their potency as local anesthetics, correlates significantly (r = 0.983) with the first-order molecular connectivity indexes. A similar linear relationship has been found[8] between the same connectivity term and solvent cavity surface areas or molecular polarizability. Both those molecular properties are important in determining the character and magnitude of biological activity. Local anesthetics are largely carbon-containing planar or aromatic molecules. In this case, in contrast to general anesthetics, the simple connectivity term ($^1\chi$) is sufficient for an adequate structural description.

Nonspecific narcotic activity, (whole animal narcosis) has been studied for 20 compounds by using their narcotic concentration against *Arenicole larvae* as a measure of narcotic potency. A fairly good correlation has been obtained[3] with $^1\chi^v$ indexes, but a close inspection of differences between experimental and calculated narcotic activity has shown that the initial set of compounds can be partitioned into two subset. The major subset of 15 compounds contains the alcohols, ethers, carbamates, a nitrile, and an amide, while the second contains two alkyl nitrates, benzene, toluene, and carbon tetrachloride. Separate correlation equations for each subset give correlation coefficients of 0.978 and 0.979, respectively. These results support the existence of two different nonspecific mechanisms of narcotic activity operating in the compounds examined. The same conclusion has also been reached with a larger set of 52 compounds of widely diverse chemical structures which exhibit different degrees of narcosis on frog tadpoles.[3] The first subset consists of nonaromatic esters, ketones, alcohols, and nitrogen-containing compounds, while the second subset is built up of aromatic and aliphatic hydrocarbons, and alkyl halides.

Quite extensive quantitative structure-activity studies of enzyme inhibitors have been accomplished by means of topological indexes[103-105] In an early study, Kier et al.[103] found a simple linear correlation between the first-order molecular connectivity index and the action of inhibitors of succinate oxidase, thymidine phosphorylase, adenosine deaminase, and butyrylcholinesterase, in their efforts to demonstrate the general applicability of connectivity indexes. More recently, two more detailed studies have been carried out on hydrazide monoamine oxidaze[104] and ribonucleotide reductase[105] inhibitors. For a set of monoamine oxidase inhibitors, the three-variable equation gave the best correlation with activity (r = 0.941). Beside the structural effects described by connectivity indexes, electronic effects, reflected through polarographic half-wave potentials, were also found to be highly important parameters in all examined correlations. To obtain more information about the character of interactions between the inhibitors and the enzyme active site, several subsets of molecules with particular substituents were examined. These observations clearly revealed that the inhibitor activity of these compounds depends on the type of ring substitution and the half-wave reduction potential of the ring, its substituents, or the ether oxygen, as well as depending on the hydrazide moiety.

The relationship between ribonucleotide reductase inhibitory potency and the structure of 28 pyridyl and benzohydroxamic acids can be described by the three-variable equation. Only three-variable equations were able to explain about 90% of the variation in inhibitory potency. Besides various connectivity indexes, Hamett values have been examined in the search for the best equation, however they have never appeared in any equation with a significant correlation. This indicates the negligible contribution of the electronic factor of ring substituents to the overall activity. The lack of an electronic contribution to the ring indicates that the whole domain of the ring takes part in the interaction with an enzyme binding site. While the interaction of the hydroxamic acid part with the metal component of the enzyme accounts for the major part of the activity, this additional interaction due to the ring and its substituent can explain the quantitative differences found in the studied compounds.

A safe, stable, and low-calory sweetener is still an actively pursued and exciting research goal. A small series of substituted nitroanilines, with a sweetness potency of up 4000 times the level of sucrose, has been studied by Hall and Kier[106] using topological indexes to establish a structure-activity relationship. Only nine substituted nitroanilines were examined. A relationship (r = 0.953) was obtained between first-order and valence first-order molecular connectivity indexes and sweetness potency. However, this result has to be used with caution and viewed as preliminary, despite the high correlation coefficient and low standard deviation, because there is a high possibility that chance correlation has been included in the equation. The hypothesis[107,108] that the chemical structure control sweetness relate three necessary (but not sufficient) structural features to taste. These three features are hydrogen bond donor, hydrogen bond acceptor, and a nonpolar region with the optimal separation between the first two being ~3Å. A serious problem in the search for noncaloric sweeteners is that both sweetness and bitternesses are usually found in a series of similar molecules. Twenty aldoximes have been used[109] to test the discriminatory power of molecular connectivity indexes (as structural descriptors) to select between sweet and bitter testing compounds. Discriminant analysis (using two indexes) properly classified nine out of ten sweet-tasting molecules and eight out of ten bitter-tasting molecules, the average percentage of correct assignment being 85%. The three incorrectly classified molecules had the lowest potencies in their taste class. This suggests that high potency is important for correct discrimination of molecules.

Connectivity indexes ($^1\chi^v$ and $^3\chi^v$) show a highly significant relationship with the bitterness threshold of amino acids, peptides, and various derivatives.[110] A qualitative discrimination of the compounds studied, based on the value of the first order valence molecular connectivity index, has been made. Three distinct classes were obtained: slightly bitter (3.9 to 5.5), bitter (5.2 to 6.6), and very bitter (6.4 to 7.7). These results may be of value in predicting the bitterness thresholds of many compounds.

Several classes of odorants (etheral odorants, floral odorants, fatty acids, and benzaldehyde-like odorants) were examined[111] by means of the molecular connectivity approach. Within each class, significant correlations were found between odor similarities to a reference standard and the connectivity term.

Hallucinogenic activity of amphetamines[112] and phenethylamines[113] has been found to correlate significantly with higher-order molecular connectivity indexes. In the important work of Kier and Hall,[112] a three-variable relationship has been found for 23 substituted amphetamines whose activity to produce an hallucinogenic effect has been expressed as concentration relative to mescaline. The best three-variable equation has been derived from a search of at least 20 terms. Despite the random number test, made by the authors, there is a great possibility[113] that chance correlation has been incorporated, at least partially, into the proposed regression equations. The relevance of the obtained regression equation has been proven by its ability to correctly predict the activity of three sets of molecules: amphetamines with undefined values of activity, a set of mescaline derivatives, and trypamines (molecules outside the phenalkylamine class). Two years later, additional experimental findings made possible more complete structure-activity studies[114] on the hallucinogenic activity of ten mescaline analogues. Although the two-variable correlation was highly significant (r = 0.97) and explained 94% of the variation in hallucinogenic activity, the variation in biological activity measurements and possibility of chance correlation (more than ten variables were screened) should be noted.

The nitrosamines, potent carcinogens,[115] known to occur in cigarette smoke, nitrate-pickled meat, and smoked fish, have been evaluated recently for their mutagenic potency using the Ames test.[116] Fifteen nitrosamines, tested under uniform conditions, were examined using molecular connectivity indexes to determine the influence of structure on this activity.[117] Two relationships of nearly equal quality have been obtained. Both equations predict activity to less than 10% of the *log* range of experimental values and show a good relationship between a molecular connectivity description of structure and mutagenic potency in the Ames test. This may be a useful start for the development of theoretical prediction of mutagenic potency in discrete classes of molecules.

Several papers[118-120] deal with the problems of the quantitative structure-activity relationships of several neurotransmitter agonists and antagonists. The muscarine receptor affinity of a large set of 104 structurally diverse acetylcholine-antagonists has been examined[118] and a satisfactory correlation has been obtained with a three-variable (connectivity indexes) equation.

An analysis of structure based on the connectivity indexes of the best regression equation emphasized the importance of onium and the bulky parts of the molecule and their mutual independence of overall activity. While the onium group contribution to experimental affinity is constant for various structural features, the contribution of bulky side chains is rather dependent on structure. The regression equation has been successfully applied in predicting the activity of other antagonists as well as of several agonist molecules. Two articles deal with the serotonin-antagonist activity of LSD analogues[119] and the serotonin-agonist activity of phenalkylamines[120] and their relationship to molecular structure expressed in terms of molecular connectivity indexes. Both sets of molecules studied have a relatively small number (16 and 17) of representatives compared with the number of variables (all zero- to sixth-order indexes) screened in the search for the best equation which avoids the risk of chance correlations. Thus, these results and the conclusions reached have to be used with great caution.

The carminative activities of 34 alcohols, esters, ethers, phenoles, and carbonyl compounds have also been examined using molecular connectivity indexes.[121] Previously, the correlation[122] of carminative activity, for the same set of compounds, against octanol-water partition coefficients (log P) explained only 50% of the variation. Inclusion of molecular connectivity terms into the correlations increased the variation explained from 50 to ~90%. Thus, it may be concluded that the steric factor is the primary cause of variation in carminative activity. This is in agreement with the mechanism proposed,[122] in which carminative activity depends on the availability of the oxygen atom in the functional group for interaction with the receptor. Activity is reduced when substituents attached to the oxygen obstruct this interaction.

First-order valence molecular connectivity indexes, together with Hückel energies of the highest occupied orbital, have been applied[54] in a quantitative structure-activity relationship study of protein (HSA) binding and mitodepressant activity of 29 isatine derivatives.[123] It was found that compounds with higher values of the $^1\chi^v$ index have a stronger affinity for binding to proteins. A parabolic relationship between mitodepressant activity and the connectivity term ($^1\chi^v$) of isatin derivatives has been obtained. The first-order valence molecular connectivity index of isatin derivative with maximal mitodepressant activity, based on the parabolic relationship, is 7.367.

Most recently, bioconcentration factors[124] and soil sorption coefficients[125] of organic pollutants (key indicators for the behavior of chemicals in the environment) have been investigated for structure-activity relationship. A parabolic relationship (r = 0.970) has been obtained[124] between the second-order valence molecular connectivity index and the bioconcentration factor of 17 chlorinated hydrocarbons. The bioconcentration factor of several commercial pesticides (DDT, DDD, DDE, dieldrin, and heptachlor) has been predicted very accurately. The same connectivity term ($^2\chi^v$) has an excellent linear relationship[125] (r = 0.99) with the soil sorption coefficient of eight polycyclic aromatic hydrocarbons and their methyl derivatives, tested under uniform conditions.[126] The soil sorption coefficients of other members of this class have been predicted.

The results of additional correlations between molecular connectivity indexes and different biological properties are summarized in Table 4.

Information theory, through the application of Shannon's equation, has been also used[131,132]

Table 4
MOLECULAR CONNECTIVITY AND BIOLOGICAL PROPERTIES

Property	Class of compounds	Indexes	r^a	Ref.
Fungus toxicity	Various	$^1\chi$	0.965	104
Cytochrome conversion	Phenols	$^1\chi^V$	0.914	106
Inhibition of *Aspergilus niger*	Benzyl alcohols	$^1\chi^V$; σ^b	0.937	106
Toxicity to the housefly	Phosphates	$^1\chi^V$; σ^b	0.975	106
Antimicrobial activity	Ethers	$^3\chi_p$; $^4\chi_{PC}^{V\,c}$	0.957	127
Antiviral activity	Benzimidazoles	$^6\chi_p^{\,d}$	0.950	127
Steroid classification	Steroids	$^0\chi^V$; $^1\chi^V$; $^n\chi^V$	—	128
Hypnotic action	Barbiturates	$^1\chi^V$	0.926	129
Antimicrobial activity	Halogen alkylphenols	$^1\chi$	0.975	130

[a] Correlation coefficient.
[b] Hamett value.
[c] PC-path/cluster.
[d] P-path.

in quantitative structure-activity correlations. Molecular negentropy, used to describe structural characteristics, has been successfully correlated with enzyme inhibition, nonspecific narcotic activity, and toxicity of various alcohols.

The comparability approach, proposed by Randić[12] has been applied to find structural similarities between benzomorphanes[133] and N-substituted 2-aminotetralines.[7] This approach differs fundamentally from the other graph-theoretical and QSAR methods which use an index (number) to describe a certain molecular structure which will quantitatively fit observed data and trends.[77,134] The basic assumption of the method is that similar molecular structures will yield similar enumeration for various graph theoretical characteristics.[78,135] The degree of similarity between molecular structures is proportional to the distance in "structure space" obtained by using these characteristics as coordinates in *n*-dimensional space. This approach can be applied whenever one has several standards with which the studied substances can be compared for structural similarity. This is the common case in medicinal and pharmaceutical chemistry.

The application of topological indexes in QSAR is still in its beginnings. Much more work is needed before the advantages and limitations of this approach are established. However, at this stage of development, several points are evident. The first one concerns biological data. The accuracy of biological measurements at present is not comparable to that of physics or chemistry. The quality of QSAR correlations will certainly increase with the improvement of standards in biological testing. However, this obstacle is minor when compared with the enormous complexity of biological systems. For example, biological activity may depend on a large number of structural (and substructural) features of a molecule. Hence when applying our limited mathematical tools to these enormously complex systems we tend to simplify grossly. Sometimes we are even tempted by the simplicity and elegance of the approach to accept a mathematically elegant formulation of something unreal rather than to accept an unsatisfying description of what is actually there.

The second point concerns the topological indexes. So far a really discriminative topological index has not been found.[1,23] Therefore, two structures with quite different biological action may have the same index. This justifies the continous search for distinctive (topological) indexes.[1,22,23,78,136] Topological indexes usually contain information about the molecule as a whole. They do not contain any information about the active part of a molecule which may be responsible for a particular biological result. Thus, they may be very useful for predicting and interpreting the origin of biological (and chemical) properties which essentially result from the bulk phase behavior. In the case of biological effects resulting from the presence of characteristic parts in a molecule, topological indexes are of limited use. Similarly, they are not great help in interpreting biological activity which is a result of a special molecular architecture, because the delicate steric features of molecules are not embodied in topological indexes.[1] New ideas are urgently needed in this area.

At the present level of development, the most used of topological indices is Randić's, connectivity index.[7] It has a number of attractive features and in many instances has exhibited a great predictive power. The connectivity index is the most tested in biological correlations and has been proposed, for example, as a criterion for water pollutants.[137] We recommend this index and its variants to those who want to carry out standard QSAR studies with topological indexes. The QSAR approach with $\chi_R(G)$ can be easily programed and the whole analysis can thus be automatized.

The information index (in different formulations) is potentially very usable[33] in QSAR schemes. Thus far it has been sparsely tested,[30] the reason being that it is by far the most complicated topological index available. However, if a convenient computer program is accessible, a QSAR study with the information index may easily be carried out.

The Balaban index also possesses very favorable features for use in QSAR (and QSPR) studies. Preliminary results in this direction are promising.[138]

REFERENCES

1. **Balaban, A. T., Chiriac, A., Motoc, I., and Simon, Z.,** *Steric Fit in Quantitative Structure-Activity Relations,* Lecture Notes in Chemistry, No. 15, Springer, Berlin, 1980, 22.
2. **Rouvray, D. H. and Balaban, A. T.,** in *Applications of Graph Theory,* Wilson, R. J. and Beineke, L. W., Eds., Academic Press, London, 1979, 177.
3. **Kier, L. B. and Hall, L. H.,** *Molecular Connectivity in Chemistry and Drug Research,* Academic, Press, New York, 1976.
4. **Sabljić, A. and Trinajstić, N.,** *Acta Pharm. Jugosl.,* 31, 189, 1982.
5. **Gutman, I. and Trinajstić, N.,** *Chem. Phys. Lett.,* 17, 535, 1972.
6. **Gutman, I., Ruščić, B., Trinajstić, N., and Wilcox, C. F., Jr.,** *J. Chem. Phys.,* 62, 3399, 1975.
7. **Randić, M.,** *J. Am. Chem. Soc.,* 97, 6609, 1975.
8. **Kier, L. B., Hall, L. H., Murray, W. J., and Randić, M.,** *J. Pharm. Sci.,* 64, 1971, 1975.
9. **Lovász, L. and Pelikán, J.,** *Period. Math. Hung.,* 3, 175, 1973.
10. **Cvetković, D. and Gutman, I.,** *Croat. Chem. Acta,* 49, 115, 1977.
11. **Harary, F.,** *Graph Theory,* Addison-Wesley, Reading, Mass., 1971, 151, second printing.
12. **Gutman, I. and Randić, M.,** *Chem. Phys. Lett.,* 47, 15, 1977.
13. **Muirhead, R. F.,** *Proc. Edinburgh Math. Soc.,* 19, 36, 1901; 24, 45, 1906.
14. **Hosoya, H.,** *Bull. Chem. Soc. Jpn.,* 44, 2332, 1971.
15. **Knop, J. V. and Trinajstić, N.,** *Int. J. Quantum Chem.,* S 14, 503, 1980.
16. **Wiener, H.,** *J. Am. Chem. Soc.,* 69, 17, 1947; 69, 2636, 1947; *J. Chem. Phys.,* 15, 766, 1947; *J. Phys. Chem.,* 52, 425, 1948; 52, 1082, 1948.
17. **Platt, J. R.,** *J. Chem. Phys.,* 15, 419, 1947; *J. Phys. Chem.,* 56, 328, 1952.
18. **Rouvray, D. H. and Crafford, B. C.,** *S. Afr. J. Sci.,* 72, 47, 1976; see also **Rouvray, D. H.,** *Math. Chem. (Mülheim/Ruhr),* 1, 125, 1975.
19. **Gordon, M. and Scantlebury, G. R.,** *Trans. Faraday Soc.,* 60, 604, 1964.
20. **Altenburg, K.,** *Kolloid-Z.,* 178, 112, 1961.
21. **Altenburg, K.,** *Z. Phys. Chem. (Leipzig),* 261, 389, 1980.
22. **Balaban, A. T.,** private communication, January 1982; **Balaban, A.,** *Chem. Phys. Lett.,* 89, 399, 1982.
23. **Bonchev, D., Mekenyan, O., and Trinajstić, N.,** *J. Comp. Chem.,* 2, 127, 1981.
24. **Smolenskii, E. A.,** *Zh. Fiz. Khim.,* 38, 1288, 1964.
25. **Hosoya, H., Murakami, M., and Gotoh, M.,** *Nat. Sci. Rep., Ochanomizu Univ.,* 24, 27, 1973.
26. **Bonchev, D. and Trinajstić, N.,** *J. Chem. Phys.,* 67, 4517, 1977.
27. **Bonchev, D. and Trinajstić, N.,** *Int. J. Quantum Chem.,* S 12, 293, 1978.

28. **Bonchev, D., Mekenyan, O., and Trinajstić, N.,** *Int. J. Quantum Chem.,* 17, 845, 1980; see also **Bonchev, D., Mekenyan, O., Knop, J. V., and Trinajstić, N.,** *Croat. Chem. Acta,* 52, 361, 1979, and *Gutman, I., Croat. Chem. Acta,* 54, 81, 1981.

29. **Mekenyan, O., Bonchev, D., and Trinajstić, N.,** *Int. J. Quantum Chem.,* 19, 929, 1981; **Mekenyan, O., Bonchev, D., and Trinajstić, N.** *Math. Chem. (Mülheim/Ruhr),* 11, 145, 1981.

30. **Bonchev, D. and Trinajstić, N.,** *Int. J. Quantum Chem.,* S 16, 463,1982.

31. **Brillouin, L.,** *Science and Information Theory,* Academic Press, New York, 1956.

32. **Shannon, C. and Weaver, W.,** *Mathematical Theory of Communication,* University of Illinois, Urbana, 1949.

33. **Bonchev, D.,** *Information Theoretic Indices For Characterization of Chemical Structures,* Research Studies Press, Chichister, in press; see also **Bonchev, D., Kamenski, D., and Kamenska, V.,** *Bull. Math. Biol.,* 38, 119, 1976; and **Bonchev, D.,** *Math. Chem. (Mülheim/Ruhr),* 7, 65, 1979.

34. **Balaban, A. T.,** *Theor. Chim. Acta,* 53, 355, 1979.

35. **Wilson, R. J.,** *Introduction to Graph Theory,* Oliver and Boyd, Edinburgh, 1972, 31.

36. **Bonchev, D., Balaban, A. T., and Mekenyan, O.,** *J. Chem. Inf. Comp. Sci.,* 20, 106, 1980.

37. **Bonchev, D., Balaban, A. T., and Randić, M.,** *Int. J. Quantum Chem.* 19, 61, 1981; erratum *Int. J. Quentum Chem.,* 22, 441, 1982.

38. **Bošnjak, N. and Trinajstić, N.,** in preparation.

39. **Bonchev, D., Mekenyan, O., Protić, G., and Trinajstić, N.,** *J. Chromatogr.,* 176, 149, 1979.

40. **Trinajstić, N., Protić, G., Švob, V., and Deur-Šiftar, D.,** *Kem. Ind. (Zagreb),* 28, 527, 1979.

41. **Bonchev, D. and Mekenyan, O.,** *Z. Naturforsch.,* 35a, 739, 1980.

42. **Bonchev, D., Mekenyan, O., and Polansky, O. E.,** *Z. Naturforsch.,* 36a, 643, 1981.

43. **Bonchev, D., Mekenyan, O., and Polansky, O. E.,** *Z. Naturforsch.,* 36a, 647, 1981.

44. **Randić, M.,** *J. Chromatogr.,* 161, 1, 1978.

45. **Kaliszan, R. and Foks, H.,** *Chromatographia,* 10, 364, 1977.

46. **Kaliszan, R.,** *Chromatographia,* 10, 529, 1977.

47. **Kaliszan, R. and Lamparczyk, H.,** *J. Chromatogr. Sci.,* 16, 246, 1978.

48. **Mc Gregor, T. R.,** *J. Chromatogr. Sci.,* 17, 314, 1979.

49. **Mc Gregor, T. R.,** *Textile Res. J.,* 49, 485, 1979.

50. **Kier, L. B.,** *J. Pharm. Sci.,* 70, 930, 1981.

51. **Hurtubise, R. J., Allen, T. W., and Iver, H. E. S.,** *J. Chromatogr.,* 235, 517, 1981.

52. **Kier, L. B. and Hall, L. H.,** *Eur. J. Med. Chem.,* 12, 307, 1977.

53. **Henry, D. R. and Block, J. H.,** *J. Med. Chem.,* 22, 465, 1979.

54. **Sabljić, A., Trinajstić, N., and Maysinger, D.,** *Acta Pharm. Jugosl.,* 31, 71, 1981.

55. **Hall, L. H. and Kier, L. B.,** *Eur. J. Med. Chem.,* 16, 399, 1981.

56. **Hermann, R. B.,** *J. Phys. Chem.,* 76, 2754, 1972.

57. **Kováts, E.,** *Z. Anal. Chem.,* 181, 351, 1961.

58. **Millership, J. S. and Woolfson, A. D.,** *J. Pharm. Pharmacol.,* 30, 483, 1978.

59. **Kier, L. B. and Hall, L. H.,** *J. Pharm. Sci.,* 68, 120, 1979.

60. **Randić, M.,** private communication (February, 1980).

61. **Hall, L. H., Kier, L. B., and Murray, W. J.,** *J. Pharm. Sci.,* 64, 1974, 1975.

62. **Kier, L. B., Murray, W. J., Randić, M., and Hall, L. H.,** *J. Pharm. Sci.,* 65, 1226, 1976.

63. **Murray, W. J., Kier, L. B., and Hall, L. H.,** *J. Pharm. Sci.,* 64, 1978, 1975.

64. **Parker, G. R.,** *J. Pharm. Sci.,* 67, 513, 1978.

65. **Gupta, S. P. and Singh, P.,** *Bull. Chem. Soc. Jpn.,* 52, 2745, 1979.

66. **Jeričević, Ž., Sabljić, A., and Trinajstić, N.,** in preparation.

67. **Hosoya, H., Kawasaki, K., and Mizutani, K.,** *Bull. Chem. Soc. Jpn.,* 45, 3415, 1972.

68. **Egloff, G., Sherman, J., and Dull, R. B.,** *J. Phys. Chem.,* 44, 730, 1940.

69. **Varshni, Y.,** *J. Ind. Chem. Soc.,* 28, 535, 1951.

70. **Nakanishi, K., Kurata, M., and Tamura, M.,** *J. Chem. Eng. Data,* 5, 210, 1960.

71. **Kreglewski, A. and Zwolinski, B. J.,** *J. Phys. Chem.,* 65, 1050, 1961.

72. **Hosoya, H.,** *J. Chem. Doc.,* 12, 181, 1972.

73. **Mizutani, K., Kawasaki, K., and Hosoya, H.,** *Nat. Sci. Rep., Ochanomizu Univ.,* 22, 39, 1971.

74. **Kawasaki, K., Mizutani, K., and Hosoya, H.,** *Nat. Sci. Rep., Ochanomizu Univ.,* 22, 181, 1971.

75. **Mekenyan, O., Bonchev, D., and Trinajstić, N.,** *Int. J. Quantum Chem.,* 18, 369, 1980.

76. **Balaban, A. T. and Motoc, I.,** *Math. Chem. (Mülheim/Ruhr),* 5, 197, 1979; **Motoc, I., Balaban, A. T., Mekenyan, O., and Bonchev, D.,** *Math. Chem.* (Mülheim/Ruhr), 13, 369,1982.

77. **Randić, M.,** *Int. J. Quantum Chem.,* S 5, 245, 1978.

78. **Randić, M. and Wilkins, C. L.,** *J. Phys. Chem.,* 83, 1525, 1979.

79. **Randić, M. and Wilkins, C. L.,** *Int. J. Quantum Chem.,* 18, 1005, 1980.

80. **Wilkins, C. L. and Randić, M.,** *Theor. Chim. Acta,* 58, 45, 1980.

81. **Edwards, J. T.,** *Can. J. Chem.,* 48, 1897, 1980.

82. **Randić, M. and Trinajstić, N.,** *Math. Chem. (Mülheim/Ruhr),* 13, 241, 1982.
83. **Crum-Brown, A. and Fraser, T.,** *Trans. R. Soc. Edinburgh,* 25, 151, 1868/69; ib., 25 693, 1868/69.
84. **Purcell, W. P., Bass, G. E., and Clayton, J. M.,** *Strategy of Drug Design,* John Wiley & Sons, New York, 1973.
85. **Martin, Y. C.,** *Quantitative Drug Design,* Marcel Dekker, New York, 1978.
86. **Sabljić, A.,** *Period. Biol.,* 81, 645, 1979.
87. **Martin, Y. C.,** *J. Med. Chem.,* 24, 229, 1981.
88. **Free, S. M., Jr. and Wilson, J. W.,** *J. Med. Chem.,* 7, 395, 1964.
89. **Hansch, C.,** *Acc. Chem. Res.,* 2, 232, 1969.
90. **Lien, E.,** *J. Am. Pharm. Educ.,* 33, 368, 1969.
91. **Pullman, B. and Pullman, A.,** *Quantum Biochemistry,* Wiley-Interscience, New York, 1963.
92. **Kier, L. B.,** *Molecular Orbital Theory in Drug Research,* Academic Press, New York, 1971.
93. **Richards, W. G.,** *Quantum Pharmacology,* Butterworths, London, 1977.
94. **Hansch, C. and Fujita, T.,** *J. Am. Chem. Soc.,* 86, 1616, 1964.
95. **Kier, L. B.,** *J. Pharm. Sci.,* 61, 1394, 1972.
96. **Agin, D., Hersch, L., and Holtzman, D.,** *Proc. Natl. Acad. Sci. U.S.A.,* 53, 952, 1965.
97. **Kier, L. B. and Hall, L. H.,** *J. Pharm. Sci.,* 70, 583, 1981.
98. **Di Paolo, T., Kier, L. B., and Hall, L. H.,** *Mol. Pharmacol.,* 13, 31, 1977.
99. **Di Paolo, T., Kier, L. B., and Hall, L. H.,** *J. Pharm. Sci.,* 68, 39, 1979.
100. **Swain, C. G. and Lupton, E. C., Jr.,** *J. Am. Chem. Soc.,* 90, 4328, 1968.
101. **Di Paolo, T.,** *J. Pharm. Sci.,* 67, 564, 1978.
102. **Di Paolo, T.,** *J. Pharm. Sci.,* 67, 566, 1978.
103. **Kier, L. B., Murray, W. J., and Hall, L. H.,** *J. Med. Chem.,* 18, 1272, 1975.
104. **Richard, A. J. and Kier, L. B.,** *J. Pharm. Sci.,* 69, 124, 1980.
105. **van't Riet, B., Kier, L. B., and Elford, H. L.,** *J. Pharm. Sci.,* 69, 586, 1980.
106. **Hall, L. H. and Kier, L. B.,** *J. Pharm. Sci.,* 66, 642, 1977.
107. **Shallenberger, R. S. and Acree, T. E.,** *Nature (London),* 216, 480, 1967.
108. **Kier, L. B.,** *J. Pharm. Sci.,* 61, 1394, 1972.
109. **Kier, L. B.,** *J. Pharm. Sci.,* 69, 416, 1980.
110. **Gardner, R. J.,** *J. Sci. Food Agric.,* 31, 23, 1980.
111. **Kier, L. B., Di Paolo, T., and Hall, L. H.,** *J. Theor. Biol.,* 67, 585, 1977.
112. **Kier, L. B. and Hall, L. H.,** *J. Med. Chem.,* 20, 1631, 1977.
113. **Topliss, J. G. and Edwards, R. P.,** *J. Med. Chem.,* 22, 1238, 1979.
114. **Glennon, R. A., Kier, L. B., and Shulgin, A. T., Jr.,** *J. Pharm. Sci.,* 68, 906, 1979.
115. **Druckey, H., Pressman, R., Cluankovic, S., and Schmall, D.,** *Z. Krebsforsch.,* 69, 103, 1967.
116. **Mc Cann, J., Choi, E., Yamasaki, E., and Ames, B. N.,** *Proc. Natl. Acad. Sci. U.S.A.,* 72, 5135, 1975.
117. **Kier, L. B., Simons, R. J., and Hall, L. H.,** *Pharm. Sci.,* 67, 725, 1978.
118. **Kier, L. B. and Hall, L. H.,** *J. Pharm. Sci.,* 67, 1408, 1978.
119. **Glennon, R. A. and Kier, L. B.,** *Eur. J. Med. Chem.,* 13, 219, 1978.
120. **Kier, L. B. and Glennon, R. A.,** *Life Sci.,* 22, 1589, 1978.
121. **Evans, B. K., James, K. C., and Luscombe, D. K.,** *J. Pharm. Sci.,* 68, 370, 1979.
122. **Evans, B. K., James, K. C., and Luscombe, D. K.,** *J. Pharm. Sci.,* 67, 277, 1978.
123. **Movrin, M., Jakševac, M., and Medić-Šarić, M.,** *Acta Pharm. Jugosl.,* 26, 67, 1976; **Maysinger, D., Medić-Šarić, M., Movrin, M., Lien, E. J., and Ban, J.,** *Eur. J. Med. Chem.,* 13, 515, 1971; **Maysinger, D., Biruš, M., and Movrin, M.,** *Acta Pharm. Jugosl.,* 30, 9, 1980.
124. **Sabljić, A. and Protić, M.,** *Chem.-Biol. Interact.,* 42, 301, 1982.
125. **Sabljić, A. and Protić, M.,** *Bull. Environ. Contam. Toxicol.,* 28, 162, 1982.
126. **Karickhoff, S. W., Brown, D. S., and Scott, D. S.,** *Water Res.,* 13, 241, 1979.
127. **Hall, L. H. and Kier, L. B.,** *J. Pharm. Sci.,* 67, 1743, 1978.
128. **Henry, D. R. and Block, J. H.,** *Eur. J. Med. Chem.,* 15, 133, 1980.
129. **Bonjean, M.-C. and Duc, C. L.,** *Eur. J. Med. Chem.,* 13, 73, 1978.
130. **Hall, L. H. and Kier, L. B.,** *Eur. J. Med. Chem.,* 13, 89, 1978.
131. **Kier, L. B.,** *J. Pharm. Sci.,* 69, 807, 1980.
132. **Roychaudhury, C., Basak, S. C., Roy, A. B., and Ghosh, J. J.,** *Indian Drugs,* December, 1, 1980.
133. **Randić, M. and Wilkins, C. L.,** *Int. J. Quantum Chem., Quantum Biol. Symp.,* 6, 55, 1979.
134. **Randić, M.,** private communication (March, 1982).
135. **Randić, M.,** *The Nature of Chemical Structure,* John Wiley & Sons, New York, in press.
136. **Kier, L. B. and Hall, L. H.,** *J. Pharm. Sci.,* 65, 1806, 1976.
137. **Kland, M. J.,** in *Water Chlorination, Environmental Impact and Health Effects,* Vol. II, Jolley, R. L., Garcher, H., and Hamilton, D. H., Jr., Eds., Ann Harbor Science, Ann Arbor, 1978, 451.
138. **Trinajstić, N.,** in progress.

Chapter 5

ISOMER ENUMERATION

The enumeration of isomeric structures is one of the oldest uses of graph theory in chemistry.[1] However, besides graph theory the basic mathematical tools necessary for isomer enumeration studies are combinatorial theory[2] and to a lesser extent group theory.[3] *Isomers* are chemical compounds possessing identical molecular formulae and molecular weights, but differing in the nature or sequence in bonding of their atoms, or in the arrangement of their atoms in space, and consequently exhibit at least some different (physical and chemical) properties.[4] The term *isomerism* and the definition of isomers was introduced by Berzelius[5] in 1830. Compounds without isomers are called *unimers*.[6]

Here we will be mostly interested in enumeration of *structural isomers*. Structural isomers have identical molecular formulae but their structures are different.[7] They were first recognized in the last century by Butlerov.[8]

There are several techniques available for the isomer enumeration of acyclic and cyclic molecules. We will discuss the enumeration techniques developed by Cayley,[9,10] Henze and Blair,[11,12] and Pólya,[13] and their applications.

I. THE CAYLEY GENERATING FUNCTIONS

Cayley was first who attempted to enumerate the isomeric alkanes $C_N H_{2N+2}$ and alkyl radicals $C_N H_{2N+1}$. He represented the carbon skeletons of alkanes by *trees* in which the maximal vertex degree is four.

Example

2,3-dimetylbutane   (alkane) tree

Similarly, alkyl radicals were depicted by *rooted trees* in which again the maximal vertex degree is four. (See discussion about the tree and rooted tree graphs in Chapter 2, Volume I, Section V).

Example

1,2,3 – trimethylpropyl   rooted tree

(Rooted) trees corresponding to (alkyl radicals) alkanes are sometimes called the Cayley (rooted) trees.[14]

The graph theoretical representation of isomeric pentanes (C_5H_{10}) and pentyl radicals (C_5H_9). by means of trees and rooted trees is shown below.

Isomeric pentane trees

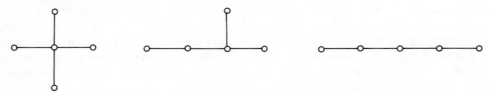

Isomeric pentyl radical trees

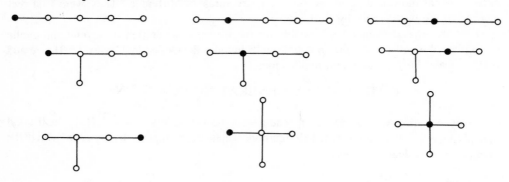

A. Enumeration of Trees

The mathematical theory of trees was introduced in the last century.[9,10,15-20] Cayley[9] first used the term *trees* in 1857, although Kirchhoff[15] first utilized the concept in his fundamental work on electrical networks in 1847.

Cayley developed a generating functions for enumeration of rooted trees,[9]

$$(1 - x)^{-1} (1 - x^2)^{-A_1} (1 - x^3)^{-A_2} \ldots = 1 + A_1 x +$$

$$A_2 x^2 + A_3 x^3 \ldots \qquad (1)$$

where x is a variable, N the number of vertices, and A_N the number of rooted trees for a given N. Enumeration of rooted trees up to N = 13 by means of Equation (1) in the Cayley way is given in Table 1.

Ten years later, Jordan[17] discovered the existence of *center* and *bicenter* of the tree (see definitions of center and bicenter in (Chapter 4, Section I.B.9). Every tree has a center or a bicenter, but not both. Thus, the trees with the center are called *centric trees,* while those trees with the bicenter are called *bicentric trees.*

Example

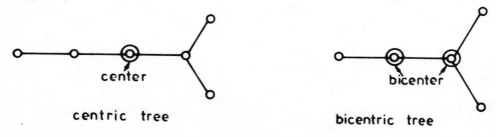

Table 1

ENUMERATION OF ROOTED TREES UP TO N = 13 BY MEANS OF THE CAYLEY GENERATING FUNCTIONS

	$(1-x^2)^{-1}$	$(1-x^3)^{-2}$	$(1-x^4)^{-4}$	$(1-x^5)^{-9}$	$(1-x^6)^{-20}$	$(1-x^7)^{-48}$	$(1-x^8)^{-115}$	$(1-x^9)^{-286}$	$(1-x^{10})^{-719}$	$(1-x^{11})^{-1842}$	$(1-x^{12})^{-4766}$	$(1-x^{13})^{-12486}$
$A_1 = 1$	(1)	—	—	—	—	—	—	—	—	—	—	—
$A_2 = 1$	(2)	—	—	—	—	—	—	—	—	—	—	—
$A_3 = 1$	2	(4)	—	—	—	—	—	—	—	—	—	—
$A_4 = 1$	3	5	(9)	—	—	—	—	—	—	—	—	—
$A_5 = 1$	3	7	11	(20)	—	—	—	—	—	—	—	—
$A_6 = 1$	4	11	19	28	(48)	—	—	—	—	—	—	—
$A_7 = 1$	4	13	29	47	67	(115)	—	—	—	—	—	—
$A_8 = 1$	5	17	47	83	123	171	(286)	—	—	—	—	—
$A_9 = 1$	5	23	61	142	222	318	433	(719)	—	—	—	—
$A_{10} = 1$	6	27	91	235	415	607	837	1123	(1842)	—	—	—
$A_{11} = 1$	6	33	125	341	741	1173	1633	2205	2924	(4766)	—	—
$A_{12} = 1$	7	42	180	531	1301	2261	3296	4440	5878	7720	(12486)	—

Table 2
ENUMERATION OF TREES UP TO N = 13 BY COUNTING THE CENTRIC AND BIOCENTRIC TREES (CAYLEY, 1875)

N	1	2	3	4	5	6	7	8	9	10	11	12	13
Number of centric trees	1	0	1	1	2	3	7	12	27	55	127	284	682
Number of bicentric trees	0	1	0	1	1	3	4	11	20	51	108	267	618
Number of trees	1	1	1	2	3	6	11	23	47	106	235	551	1301

Cayley made use of the work by Jordan and obtained formulae for separated enumeration of centric and bicentric trees.[19] The sum of centric and bicentric trees for a given N produces the total number of isomeric trees with N vertices. These results are given in Table 2.

Later (in 1881) Cayley improved his method for enumerating trees by using the generating functions and already known A_N numbers from the enumeration of rooted trees.[20] Thus, the trees T_N can be counted by means of the following formulae,

$$T_1 = 1 \tag{2}$$

$$T_2 = (1/2) A_0 (A_0 + 1) \tag{3}$$

$$T_3 = \text{the coefficient at } x^2 \text{ in } (1 - x)^{-A_0} \tag{4}$$

$$T_4 = (1/2) A_2 (A_2 + 1) + \text{the coefficient at } x^3 \text{ in } (1 - x)^{-A_0} \tag{5}$$

$$T_5 = \text{the coefficient at } x^4 \text{ in } (1 - x)^{-A_0} (1 - x^2)^{-A_1} \tag{6}$$

$$T_6 = (1/2) A_3 (A_3 + 1) + \text{the coefficient at } x^5 \text{ in } (1 - x)^{-A_0}$$
$$(1 - x^2)^{-A_1} \tag{7}$$

$$T_7 = \text{the coefficient at } x^6 \text{ in } (1 - x)^{-A_0} (1 - x^2)^{-A_1}$$
$$(1 - x^3)^{-A_2} \tag{8}$$

$$T_8 = (1/2) A_4 (A_4 + 1) + \text{the coefficient at } x^7 \text{ in } (1 - x)^{-A_0}$$
$$(1 - x^2)^{-A_1} (1 - x^3)^{-A_3} \tag{9}$$

$$T_9 = \text{the coefficient at } x^8 \text{ in } (1 - x)^{-A_0} (1 - x^2)^{-A_1}$$
$$(1 - x^3)^{-A_2} \cdot (1 - x^4)^{-A_3} \tag{10}$$

Example

Enumeration of trees for N = 7

$$(1 - x)^{-A_0} = 1 + A_0 x + (1/2) A_0 (A_0 + 1) x^2 + (1/6) A_0 (A_0 + 1) (A_0 + 2) x^3$$
$$+ (1/24) A_0 (A_0 + 1) (A_0 + 2) (A_0 + 3) x^4 + (1/120) A_0 (A_0 + 1)$$
$$(A_0 + 2) (A_0 + 3) (A_0 + 4) x^5 + (1/720) A_0 (A_0 + 1) (A_0 + 2)$$
$$(A_0 + 3) (A_0 + 4) (A_0 + 5) x^6 + \ldots$$

$$(1 - x^2)^{-A_1} = 1 + A_1 x^2 + (1/2) A_1 (A_1 + 1) x^4 + (1/6) A_1 (A_1 + 1) (A_1 + 2) x^6 + \ldots$$

$$(1 - x^3)^{-A_2} = 1 + A_2 x^3 + (1/2) A_2 (A_2 + 1) x^6 + \ldots$$

$$T_7 = (1/720) A_0 (A_0 + 1) (A_0 + 2) (A_0 + 3) (A_0 + 4) (A_0 + 5) +$$

$$(1/6) A_1 (A_1 + 1) (A_1 + 2) + (1/2) A_2 (A_2 + 1) + A_0 A_1 A_2 +$$

$$(1/4) A_0 A_1 (A_0 + 1) (A_1 + 1) + (1/6) A_0 A_2 (A_0 + 1) (A_0 + 2) +$$

$$(1/24) A_0 A_1 (A_0 + 1) (A_0 + 2) (A_0 + 3) = 11$$

The use of the Cayley approach is rather tedious and the error could be easily made. Actually Cayley made several errors in his work.[1] Cayley was very close to derive a general formula for the number of trees in terms of the number of rooted trees, entirely by means of generating functions, but did not find it. However, half a century later Otter[21] derived an elegant formula for counting trees in terms of rooted trees using the Cayley generating functions,

$$T(x) = A(x) - (1/2) [A^2(x) - A(x^2)] \tag{11}$$

Example

Enumeration of trees by means of Equation (11)

$$A(x) = x + x^2 + 2x^3 + 4x^4 + 9x^5 + 20x^6 + 48x^7 + \ldots$$

$$A^2(x) = x^2 + 2x^3 + 5x^4 + 12x^5 + 30x^6 + 74x^7 + \ldots$$

$$A(x^2) = x^2 + x^4 + 2x^6 + 4x^8 + \ldots$$

$$(1/2) [A^2(x) - A(x^2)] = x^3 + 2x^4 + 6x^5 + 14x^6 + 37x^7 + \ldots$$

$$T(x) = x + x^2 + x^3 + 2x^4 + 3x^5 + 6x^6 + 11x^7 + \ldots$$

For the reader's information we give in Table 3 the number of distinct trees and rooted trees up to 50 vertices obtained by the computer count of N-tuples.[22] (In this work[22] trees and rooted trees with N vertices are represented by N tuples of nonnegative integers smaller than N). One clearly sees that the number of isomeric (rooted) trees increases rapidly with the increase of N. This is indicative that the method of Cayley was on rather limited practical value for large N.

B. Enumeration of Alkanes

Cayley[10] was the first to realize the potentiality of the mathematical theory of trees for the enumeration of (hydrocarbon) isomers. He enumerated alkanes[23] and alkyl radicals[24] up to the 13th member, but the number of isomers obtained for C_{12} and C_{13} alkanes (357 and 799), and C_{13} alkyl radicals (7638) were incorrect. In addition, he stated that no compact formula could be found for the isomer enumeration of alkanes. However, inspite of short-comings, the work of Cayley had a considerable impact on (theoretical and mathematical) chemists of his time. Almost immediately after Cayley's paper on enumeration of alkanes,[23] a work by Schiff[25] appeared in which he correctly counted distinct alkanes, alkenes, and alkyl radicals up to N = 10. Schiff also attempted to calculate the number of isomeric dodecanes, $C_{12}H_{26}$, obtaining the same erroneous values as that Cayley. The errors in computing the number of C_{12} and C_{13} alkane isomers were first corrected (C_{12}: 355 and C_{13}: 802) by Herrmann[26] about 5 years later. It is interesting to note that Losanitsch[27] was arguing that the number of C_{12} isomeric alkanes is neither 357 nor 355, but 354. There was quite

Table 3
THE NUMBER OF TREES AND ROOTED TREES WITH N VERTICES

N	Number of trees	Number of rooted trees
1	1	1
2	1	1
3	1	2
4	2	4
5	3	9
6	6	20
7	11	48
8	23	115
9	47	286
10	106	719
11	235	1842
12	551	4766
13	1301	12486
14	3159	32973
15	7741	87811
16	19320	235381
17	48629	634847
18	123867	1721159
19	317955	4688676
20	823065	12826228
21	2144505	35221832
22	5623756	97055181
23	14828074	268282855
24	39299897	743724984
25	104636890	2067174645
26	279793450	5759636510
27	751065460	16083734329
28	2023443032	45007066269
29	5469566585	126186554308
30	14830871802	354426847597
31	40330829030	997171512998
32	109972410221	2809934352700
33	300628862480	7929819784355
34	823779631721	22409533673568
35	2262366343746	63411730258053
36	6226306037178	179655930440464
37	17169677490714	509588049810620
38	47436313524262	1447023384581029
39	131290543779126	4113254119923150
40	363990257783343	11703780079612453
41	1010748076717151	33333125878283632
42	2810986483493475	95020085893954917
43	7828986221515605	271097737169671824
44	21835027912963086	774088023431472074
45	60978390985918906	2212039245722726118
46	170508699155987862	6325843306177425928
47	477355090753926460	18103111141539779470
48	1337946100045842285	51842285219378800562
49	3754194185716399992	148558992149369434381
50	10545233702911509534	425976989835141038353

a discussion between Herrmann[28,30] and Losanitsch[29] at the end of the last century about whose number is correct. Herrmann, of course, produced a correct value (355).[24] However, none of the above authors, including several others,[31-33] were able to produce a reliable technique for enumeration of alkanes with N large. Their methods have been unwieldy and often led to incorrect results. The first significant advance in this area after Cayley came in 1931 when Henze and Blair,[11,12,34,45] at the University of Texas at Austin, developed the recursion formulae for enumeration of alkanes and related structures.

II. THE HENZE-BLAIR APPROACH

The Henze-Blair approach[11,12,34,35] is based on three suppositions: (1) Cayley's conclusions that no formula could be derived for isomer counts for members of homologous series: (2) free rotation about single carbon-carbon bonds does not bring forth new isomeric structures, and (3) if the isomer enumeration for a given member of an homologous series having N atoms could be established, a germane recursion formula would produce the isomer count for the next member having N + 1 atoms.

The essence of the Henze-Blair approach[11] will be illustrated for the enumeration of alcohols, $C_N H_{2N+1} \cdot OH$. We will represent alcohols by the rooted trees. In the rooted trees we differ the *primary root* (a rooted vertex with valency one), the *secondary root* (a rooted vertex with valency two), the *tertiary root* (a rooted vertex with a valency three), and the *quartenary root* (a rooted vertex with a valency four). In the case of alcohols, the rooted tree(s) with the primary (secondary, tertiary) root would represent the primary (secondary, tertiary) alcohol.

Example

1-hydroxy-3-metylpentane
(primary alcohol)

2-hydroxy-3-methylpentane
(secondary alcohol)

3-hydroxy-3-methylpentane
(tertiary alcohol)

rooted tree with the
primary root

rooted tree with the
secondary root

rooted tree with the
tertiary root

The number of rooted trees with the primary root, secondary root, and tertiary root containing N vertices are denoted by p_N, s_N, and t_N, respectively. The total number of isomeric rooted trees (e.g., alcohols) L_N is then given by,

$$L_N = p_N + s_N + t_N \tag{12}$$

The replacement of an $-OH$ group in any alcohol by a CH_2OH group leads always to formation of a primary alcohol. Thus, the number of primary alcohols p_N is given by,

$$p_N = L_{N-1} \tag{13}$$

The secondary alcohol may be imagined to consist of two alkyl radical parts R_i and R_j connected to the $\geq$CH·OH group. The number of carbon atoms (i and j) in R_i and R_j is obviously $N - 1$, i.e., $i + j = N - 1$. This may be utilized for deriving equations for s_N,

$$s_N = \begin{cases} L_1 \cdot L_{N-2} + L_2 \cdot L_{N-3} + \ldots + L_{(N-2)/2} \cdot L_{N/2} & ; N = \text{even} \\[2mm] L_1 \cdot L_{N-2} + L_2 \cdot L_{N-3} + \ldots + L_{(N-3)/2} \cdot L_{(N+1)/2} + \\[2mm] (1/2) L_{(N-1)/2} \left[1 + L_{(N-1)/2}\right] & ; N = \text{odd} \end{cases}$$

(14)

In the case of tertiary alcohols formation of a molecule may be imagined to arise from joining three alkyl radicals R_i, R_j, and R_k to the $\geq$C·OH group. The number of carbon atoms (i, j, and k) in R_i, R_j, and R_k is again $N - 1$, i.e., $i + j + k = N - 1$. The equations for t_N are then given by.

$$t_N = \begin{cases} \Sigma L_i L_j L_k \; ; R_i \neq R_j \neq R_k \; ; i + j + k = N - 1 \\[2mm] (1/2) \Sigma L_i (1 + L_i) L_j \; ; R_i = R_j \neq R_k \; ; 2i + k = N - 1 \\[2mm] (1/6) L_i (1 + L_i) (2 + L_i) \; ; R_i = R_j = R_k \; ; 3i = N - 1 \end{cases}$$

(15)

Example

Enumeration of isomeric alcohols $C_N H_{2N+1}$·OH up to $N = 6$

(i) $p_1 = L_0$, $L_0 = 1$ (by definition)

$s_1 = 0$

$t_1 = 0$

$L_1 = p_1 + s_1 + t_1 = 1$

(ii) $p_2 = L_1 = 1$

$s_2 = 0$

$t_2 = 0$

$L_2 = p_2 + s_2 + t_2 = 1$

(iii) $p_3 = L_2 = 1$

$s_3 = (1/2) L_1 (1 + L_1) = 1$

$t_3 = 0$

$L_3 = p_3 + s_3 + t_3 = 2$

(iv) $p_4 = L_3 = 2$

$s_4 = L_1 \cdot L_2 = 1$

$t_4 = (1/6) L_1 (1 + L_1) (2 + L_1) = 1$

$L_4 = p_4 + s_4 + t_4 = 4$

(v) $p_5 = L_4 = 4$

$s_5 = L_1 \cdot L_3 + (1/2) L_2 \cdot (1 + L_2) = 3$

$t_5 = (1/2) L_1 (1 + L_1) L_2 = 1$

$L_5 = p_5 + s_5 + t_5 = 8$

(vi) $p_6 = L_5 = 8$

$s_6 = L_1 L_4 + L_2 L_3 = 6$

$t_6 = (1/2) L_1 (1 + L_1) L_3 + L_2 (1 + L_2) L_1 = 3$

$L_6 = p_6 + s_6 + t_6 = 17$

Henze and Blair calculated correctly the number of primary, secondary, and tertiary alcohols up to N = 20. Then, they extended their approach to alkanes.[11] They separated alkanes into classes according to whether the number of carbon atoms in them is *even* or *odd*. Each class is further divided into two groups. Alkanes with N = even are partitioned into *Group A'* containing those alkanes whose graphs may be divided into two constituent parts (alkyl radicals) with N/2 atoms each and *Group B'* containing those alkanes whose graphs cannot be divided into two equal parts. Alkanes with N = odd are similarly partitioned into *Group A"* containing those alkanes whose graphs may be divided into two constituents parts (alkyl radicals), one with (N + 1)/2 carbons and the other with (N − 1)/2 carbons, respectively, and *Group B"* containing remaining alkanes with N = odd whose graphs cannot be divided in the above way.

The isomers of alkanes in Group A' may be calculated using the following equation,

$$(1/2) L_{N/2} (1 + L_{N/2}) \qquad (16)$$

where the previously determined[11] values of L_N are utilized. The isomers of alkanes in Group A" may be calculated by means of the formula,

$$(1/2) L_{(N-1)/2} (1 + 2L_{(N+1)/2} - L_{(N-1)/2}) \qquad (17)$$

The isomeric alkanes in Group B' and Group B" are of two types: (1) those in which *three* branches are attached to the specified carbon atom, and (2) those in which *four* branches are connected to the specified carbon atom. The total number of isomers in Group B' and in Group B" is obtained by summing up isomers of both types.

Type (1) consists of *three* possible cases:

Case 1 — All three branches are of the different length,

$$\Sigma L_i L_j L_k \qquad (18)$$

where $i + j + k = N - 1$ and $i > j > k$. For N = even $i \leq (N/2) - 1$, and for N = odd $i \leq (N - 3)/2$.

Case 2 — Two branches are of the same length and different from the third,

$$(1/2) \Sigma L_i L_j (1 + L_i) \qquad (19)$$

where $2i + j = N - 1$. For N = even $i, j \leq (N/2) - 1$, and for N = odd $i, j \leq (N - 3)/2$.

Case 3 — All three branches are the same size,

$$(1/6) \sum L_i \, (1 + L_i)(1 + L_i) \tag{20}$$

where $3i = N - 1$.

Type (2) consists of *five* possible cases.

Case 1 — All four branches are of different length,

$$\sum L_h \, L_i \, L_j \, L_k \tag{21}$$

where $h + i + j + k = N - 1$ and $h > i > j > k$. For $N = $ even $h \leq (N/2) - 1$, and for $N = $ odd $h = (N - 3)/2$.

Case 2 — Two branches are of the same length while each of the two others is of the different length,

$$(1/2) \sum L_i \, L_j \, L_k \, (1 + L_k) \tag{22}$$

where $2\,i + j + k = N - 1$ and $i > j > k$.

Case 3 — Three branches are of the same length and different from the fourth,

$$(1/6) \sum L_i \, L_j \, (1 + L_i)(2 + L_i) \tag{23}$$

where $3\,i + j = N - 1$ and $i > j$.

Case 4 — All four branches are of the same length,

$$(1/24) \sum L_i \, (1 + L_i)(2 + L_i)(3 + L_i) \tag{24}$$

where $4\,i = N - 1$.

Case 5 — Four branches are partitioned into two sets containing two branches each. The individual members of each set are of the same length and different from the members of the other set.

$$(1/4) \sum L_i \, L_j \, (1 + L_i)(1 + L_j) \tag{25}$$

where $2\,i + 2\,j = N - 1$ and $i > j$.

The total number of structural isomers of alkanes with N carbon atoms may be determined by adding to the number of isomers calculated in Group A' (Group A''), the number of isomers calculated in each of the cases of Group B' (Group B'').

Example

Enumeration of isomeric dodecanes $C_{12}H_{24}$

$$N = 12$$

<u>Group A′</u>

$$N/2 = 6$$

$$(1/2) \cdot L_6 \cdot (1 + L_6) = (1/2) \cdot 17 \cdot (1 + 17) = 153$$

<u>Group B′</u>

$$N - 1 = 11; \ (N/2) - 1 = 5$$

Type (i): **Case 1**

$$L_5 \cdot L_4 \cdot L_2 = 8 \cdot 4 \cdot 1 = 32$$

Case 2

$$(1/2) \cdot L_5 \cdot L_1 \cdot (1 + L_5) = (1/2) \cdot 8 \cdot 1 \cdot (1 + 8) = 36$$

$$(1/2) \cdot L_4 \cdot L_3 \cdot (1 + L_4) = (1/2) \cdot 4 \cdot 2 \cdot (1 + 4) = 20$$

$$(1/2) \cdot L_3 \cdot L_5 \cdot (1 + L_3) = (1/2) \cdot 2 \cdot 8 \cdot (1 + 2) = \underline{24}$$

$$80$$

Type (ii): **Case 1**

$$L_5 \cdot L_3 \cdot L_2 \cdot L_1 = 8 \cdot 2 \cdot 1 \cdot 1 = 16$$

Case 2

$$(1/2) \cdot L_4 \cdot L_2 \cdot L_1 \cdot (1 + L_4) = (1/2) \cdot 4 \cdot 1 \cdot 1 \cdot (1 + 4) = 10$$

$$(1/2) \cdot L_3 \cdot L_4 \cdot L_1 \cdot (1 + L_3) = (1/2) \cdot 2 \cdot 4 \cdot 1 \cdot (1 + 2) = 12$$

$$(1/2) \cdot L_2 \cdot L_4 \cdot L_3 \cdot (1 + L_2) = (1/2) \cdot 1 \cdot 4 \cdot 2 \cdot (1 + 1) = 8$$

$$(1/2) \cdot L_1 \cdot L_5 \cdot L_4 \cdot (1 + L_1) = (1/2) \cdot 1 \cdot 8 \cdot 4 \cdot (1 + 1) = \underline{32}$$

$$62$$

Case 3

$$(1/6) \cdot L_3 \cdot L_2 \cdot (1 + L_3) \cdot (2 + L_3) = (1/6) \cdot 2 \cdot 1 \cdot (1 + 2) \cdot (2 + 2) = 4$$

$$(1/6) \cdot L_2 \cdot L_5 \cdot (1 + L_2) \cdot (2 + L_2) = (1/6) \cdot 1 \cdot 8 \cdot (1 + 1) \cdot (2 + 1) = \underline{8}$$

$$12$$

The total number of isomeric dodecanes is $T_{12} = 355$.

Henze and Blair[12] calculated the number of alkane isomers up to $N = 20$, but have also listed the number of structural isomers for $C_{25}H_{52}$, $C_{30}H_{62}$, and $C_{40}H_{82}$ alkanes. A year later, Perry[36] augmented their work by producing the numbers of isomeric alcohols from $N = 21$

to N = 30 and the numbers of isomeric alkanes from N = 21 to N = 39 and all isomers with common formula $C_{60}H_{122}$. During this work he detected only one numerical error in the work by Henze and Blair.[12] They reported erroneously the number of isomers of $C_{19}H_{40}$ to be 147284 instead of 148284.

Although the application of the Henze-Blair approach involves tedious calculations, it has been used a lot for the enumeration of structural isomers and stereoisomers of various classes of compounds (saturated hydrocarbons, unsaturated hydrocarbons, acetylenes, alcohols, etc.).[37-42] Actually some 40 years after the introduction, the computer program was written[43] based on the Henze-Blair method and computation was made for isomeric alkanes up to N = 50.

For the reader's information, we give in Table 4 the number of alkanes C_NH_{2N+2}, the number of primary alcohols $C_{N-1}H_{2N-1}CH_2OH$, the secondary alcohols $C_{N-1}H_{2N}CHOH$, the tertiary alcohols $C_{N-1}H_{2N}COH$, and the total number of alcohols up to N = 50 obtained by the computer count based on the N-tuple representation of trees and rooted trees.[22]

III. THE PÓLYA ENUMERATION METHOD

Pólya proposed in 1937 the most powerful enumeration method that is available to the chemist.[5,13,44-50] Although the method rests on some results by Redfield,[51] and Lunn and Senior,[52] it gave the first recipe for the systematic derivation of counting series by making a collective use of symmetry properties of molecules, generating functions, and weighting factors.

The counting series is given as,

$$S(x) = \sum_{n=0}^{\infty} a_n x^n \tag{26}$$

where n characterizes all the isomers corresponding to the same molecular weight, a_n stands for the number of these isomers. However, the *cycle index* plays an important role in the derivation of the enumeration series. It is connected with the symmetry characteristics of a molecule and for every symmetry operation there is a corresponding cycle index. The cycle index, Z(H), is defined by,

$$Z(H) = (1/h) \sum{}' n_{i_1 i_2 \ldots i_p} f_1^{i_1} f_2^{i_2} \ldots f_p^{i_p} \tag{27}$$

where Z(H) is the cycle index for a permutation group H of order h, p is the number of vertices permuted, the $f_1, f_2, \ldots, f_p$ are variables, while $n_{i_1 i_2 \ldots i_p}$ represents the number of permutations of H which consist of i_1 cycles of order one, i_2 cycles of order two, etc. The prime over the summation indicates that the sum is over all the set $\{i_p\}$, provided that the condition,

$$\sum_{l=1}^{p} h_i = P \tag{28}$$

always holds, where P is the total number of permutations. Let consider, as an example, the derivation of the cycle index for each symmetry operation on the benzene ring. (Benzene belongs to the point group symmetry D_{6h}).[53] This is shown in Table 5.

The resultant cycle index of benzene is obtained by addition of all cycle indices corresponding to individual symmetry operations and by dividing the sum with the order of the group,

$$Z(\text{benzene}) = (1/12) (f_1^6 + 4 f_2^3 + 3 f_1^2 f_2^2 + 2 f_3^2 + 2 f_6^1) \tag{29}$$

Table 4
THE NUMBER OF ISOMERIC ALKANES AND ALCOHOLS WITH N CARBON ATOMS

N	Number of alkanes	Number of primary alcohols	Number of secondary alcohols	Number of tertiary alcohols	Total number of alcohols
1	1	1	0	0	1
2	1	1	0	0	1
3	1	1	1	0	2
4	2	2	1	1	4
5	3	4	3	1	8
6	5	8	6	3	17
7	9	17	15	7	39
8	18	39	33	17	89
9	35	89	82	40	211
10	75	211	194	102	507
11	159	507	482	249	1238
12	355	1238	1188	631	3057
13	802	3057	2988	1594	7639
14	1858	7639	7528	4074	19241
15	4347	19241	19181	10443	48865
16	10359	48865	49060	26981	124906
17	24894	124906	126369	69923	321198
18	60523	321198	326863	182158	830219
19	148284	830219	849650	476141	2156010
20	366319	2156010	2216862	1249237	5622109
21	910726	5622109	5806256	3287448	14715813
22	2278658	14715813	15256265	8677074	38649152
23	5731580	38649152	40210657	22962118	101821927
24	14490245	101821927	106273050	60915508	269010485
25	36797588	269010485	281593237	161962845	712566567
26	93839412	712566567	747890675	431536102	1891993344
27	240215803	1891993344	1990689459	1152022025	5034704828
28	617105614	5034704828	5309397294	3081015684	13425117806
29	1590507121	13425117806	14187485959	8253947104	35866550869
30	4111846763	35866550869	37977600390	22147214029	95991365288
31	10660307791	95991365288	101827024251	59514474967	257332864506

Table 4 (continued)
THE NUMBER OF ISOMERIC ALKANES AND ALCOHOLS WITH N CARBON ATOMS

N	Number of alkanes	Number of primary alcohols	Number of secondary alcohols	Number of tertiary alcohols	Total number of alcohols
32	27711253769	257332864506	273442837014	160152652585	690928354105
33	72214088660	690928354105	735356029184	431536968270	1857821351559
34	188626236139	1857821351559	1980245349791	1164238905803	5002305607153
35	493782952902	5002305607153	5339453162253	3144681306263	13486440075669
36	1295297588128	13486440075669	14414507646239	8503434708370	36404382430278
37	3404449780161	36404382430278	38958262395690	23018134344315	98380779170283
38	8964747474595	98380779170283	105407071465709	62370701364485	266158552000477
39	23647478933969	266158552000477	285486823673472	169162601157498	720807976831447
40	62481801147341	720807976831447	773973501324306	459220572506066	1954002050661819
41	165351455535782	1954002050661819	2100240521050067	1247708120305177	5301950692017063
42	438242897692226	5301950692017063	5704211125099465	3392829794022689	14398991611139217
43	1163169707886427	14398991611139217	15505573111151541	9233204029174994	39137768751465752
44	3091461011836856	39137768751465752	42182179142471817	25146006764593896	106465954658531465
45	8227162377221203	106465954658531465	114842744354613849	68532690093294099	289841389106439413
46	21921834086683418	289841389106439413	312893758322611809	186906970120044539	789642117549095761
47	58481806621987010	789642117549095761	853091050332332960	510081778090226835	2152814945971655556
48	156192366474590639	2152814945971655556	2327481497654824913	1392929364734851485	5873225808361331954
49	4177612400765382272	5873225808361331954	6354149922621658161	3806119516574048959	16033495247557039074
50	1117743651746953270	16033495247557039074	17357929363276213245	10406136061618401441	43797554941937577760

Table 5

A PRESENTATION OF THE CYCLE INDEX CORRESPONDING TO EACH SYMMETRY OPERATION WHICH MAY BE PERFORMED ON THE BENZENE RING

Symmetry operation	Vertex interchange	Permutation grouping	Cycle index
E	1 2 3 4 5 6 1 2 3 4 5 6	1 2 3 4 5 6 1 2 3 4 5 6	f_1^6
C_6^+	1 2 3 4 5 6 6 1 2 3 4 5	1 2 3 4 5 6 6 1 2 3 4 5	f_6^1
C_6^-	1 2 3 4 5 6 2 3 4 5 6 1	1 2 3 4 5 6 2 3 4 5 6 1	f_6^1
C_3^+	1 2 3 4 5 6 5 6 1 2 3 4	1 3 5 2 4 6 5 1 3 6 2 4	f_3^2
C_3^-	1 2 3 4 5 6 3 4 5 6 1 2	1 3 5 2 4 6 3 5 1 4 6 2	f_3^2
C_2	1 2 3 4 5 6 4 5 6 1 2 3	1 4 2 5 3 6 4 1 5 2 6 3	f_2^3
$\sigma_v^{(1)}$	1 2 3 4 5 6 1 6 5 4 3 2	1 4 3 5 2 6 1 4 5 3 6 2	$f_1^2 f_2^2$
$\sigma_v^{(2)}$	1 2 3 4 5 6 5 4 3 2 1 6	3 6 1 5 2 4 3 6 5 1 4 2	$f_1^2 f_2^2$
$\sigma_v^{(3)}$	1 2 3 4 5 6 3 2 1 6 5 4	2 5 1 3 4 6 2 5 3 1 6 4	$f_2^2 f_2^2$
$\sigma_v^{(4)}$	1 2 3 4 5 6 6 5 4 3 2 1	1 6 2 5 3 4 6 1 5 2 4 3	f_2^3
$\sigma_v^{(5)}$	1 2 3 4 5 6 2 1 6 5 4 3	1 2 3 6 4 5 2 1 6 3 5 4	f_2^3
$\sigma_v^{(6)}$	1 2 3 4 5 6 4 3 2 1 6 5	1 4 2 3 5 6 4 1 3 2 6 5	f_2^3

$$Z(\text{benzene}) = (1/12)\ f_1^6 + 4f_2^3 + 3f_1^2 f_2^2 + 2f_3^2 + 2f_6^1$$

In order to obtain the number of structural isomers of benzene when k hydrogen atoms are substituted by monovalent radical R ($k = 1,2,3,4,5$ or 6) for each of the component cycle indices f_k a substitution of the type,

$$f_k^\ell = (1 + x^k)^\ell \qquad (30)$$

is made in Equation (29). The reason for this substitution is the creation of the power series in x in which the coefficient at x^k in the expansion gives directly the number of isomers which can be formed when k hydrogen atoms in the benzene are substituted by univalent atoms or groups. The substitution of (30) into (29) yields the counting polynomial,

$$Z(\text{benzene}) = 1 + x + 3x^2 + 3x^3 + 3x^4 + x^5 + x^6 \qquad (31)$$

The exponent of each term gives the number of monovalent R atoms substituted into the benzene nucleus, while the coefficients give the corresponding number of isomers. For example, since the coefficient at x^3 is *three*, it means that trisubstitution of benzene by the univalent radical or group R will produce *three* structural isomers $C_6H_3R_3$. All possible structural isomers of benzene counted by Equation (31) are presented in Figure 1.

The cycle indexes and the corresponding counting polynominals for several benzenoid hydrocarbons are given in Tables 6 and 7.

We shall now investigate how many structural isomers can be obtained if two different

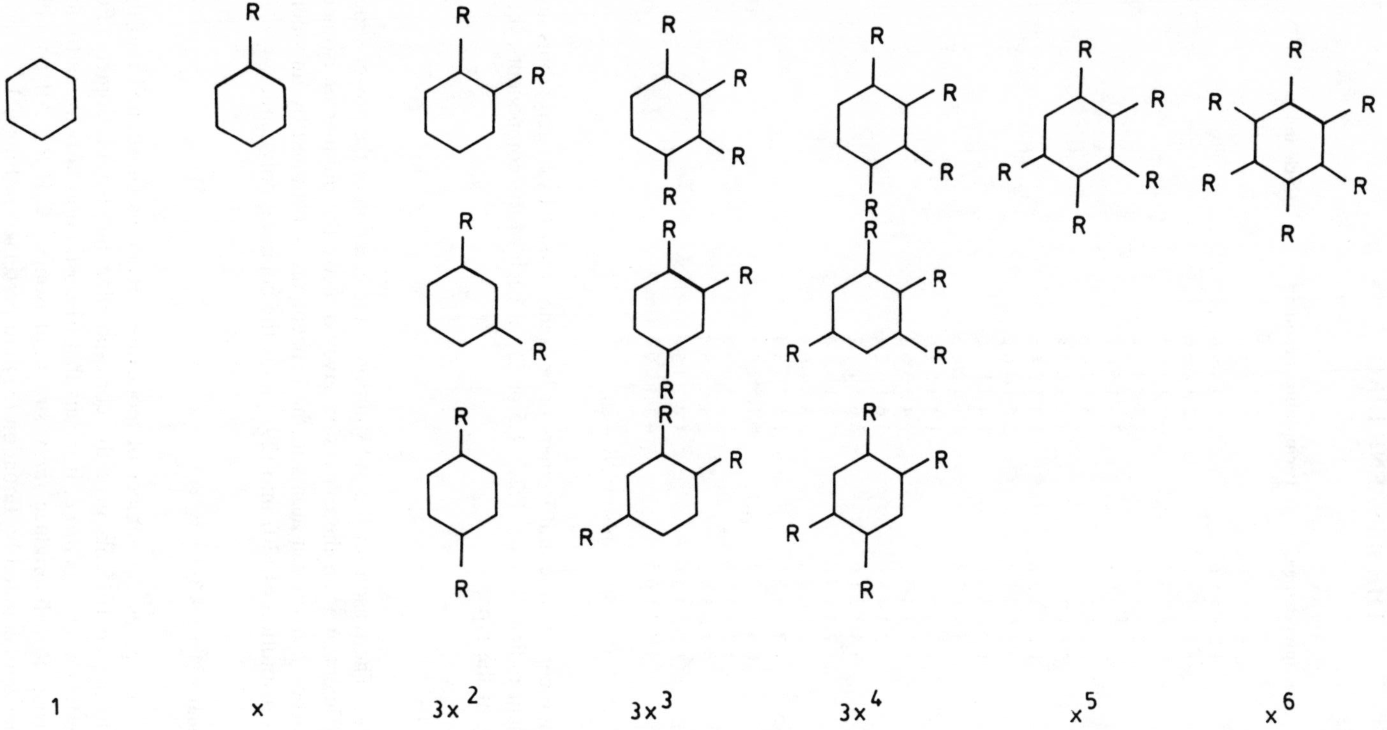

FIGURE 1.   All structural isomers formed when the hydrogen atoms in benzene are successively substituted by monovalent radicals R.

substituents enter the cyclobutadiene nucleus. In doing that we proceed along the following steps:

Calculation of the cycle index of cyclobutadiene (cyclobutadiene belongs to the point group symmetry D_{4h}).

G

Symetry operation	Vertex exchange	Permutation groupings	Cycle index
E	$\begin{pmatrix} 1 & 2 & 3 & 4 \\ 1 & 2 & 3 & 4 \end{pmatrix}$	$\begin{pmatrix} 1 \\ 1 \end{pmatrix} \begin{pmatrix} 2 \\ 2 \end{pmatrix} \begin{pmatrix} 3 \\ 3 \end{pmatrix} \begin{pmatrix} 4 \\ 4 \end{pmatrix}$	f_1^4
C_4^+	$\begin{pmatrix} 1 & 2 & 3 & 4 \\ 4 & 1 & 2 & 3 \end{pmatrix}$	$\begin{pmatrix} 1 & 2 & 3 & 4 \\ 4 & 1 & 2 & 3 \end{pmatrix}$	f_4^1
C_4^-	$\begin{pmatrix} 1 & 2 & 3 & 4 \\ 2 & 3 & 4 & 1 \end{pmatrix}$	$\begin{pmatrix} 1 & 2 & 3 & 4 \\ 2 & 3 & 4 & 1 \end{pmatrix}$	f_4^1
C_2	$\begin{pmatrix} 1 & 2 & 3 & 4 \\ 3 & 4 & 1 & 2 \end{pmatrix}$	$\begin{pmatrix} 1 & 3 \\ 3 & 1 \end{pmatrix} \begin{pmatrix} 2 & 4 \\ 4 & 2 \end{pmatrix}$	f_2^2
$\sigma_v^{(1)}$	$\begin{pmatrix} 1 & 2 & 3 & 4 \\ 1 & 4 & 3 & 2 \end{pmatrix}$	$\begin{pmatrix} 1 \\ 1 \end{pmatrix} \begin{pmatrix} 3 \\ 3 \end{pmatrix} \begin{pmatrix} 2 & 4 \\ 4 & 2 \end{pmatrix}$	$f_1^2 f_2^1$
$\sigma_v^{(2)}$	$\begin{pmatrix} 1 & 2 & 3 & 4 \\ 3 & 2 & 1 & 4 \end{pmatrix}$	$\begin{pmatrix} 2 \\ 2 \end{pmatrix} \begin{pmatrix} 4 \\ 4 \end{pmatrix} \begin{pmatrix} 1 & 3 \\ 3 & 1 \end{pmatrix}$	$f_1^2 f_2^1$
$\sigma_v^{(3)}$	$\begin{pmatrix} 1 & 2 & 3 & 4 \\ 2 & 1 & 4 & 3 \end{pmatrix}$	$\begin{pmatrix} 1 & 2 \\ 2 & 1 \end{pmatrix} \begin{pmatrix} 3 & 4 \\ 4 & 3 \end{pmatrix}$	f_2^2
$\sigma_v^{(4)}$	$\begin{pmatrix} 1 & 2 & 3 & 4 \\ 4 & 3 & 2 & 1 \end{pmatrix}$	$\begin{pmatrix} 1 & 4 \\ 4 & 1 \end{pmatrix} \begin{pmatrix} 2 & 3 \\ 3 & 2 \end{pmatrix}$	f_2^2

$$Z(\text{cyclobutadiene}) = (1/8)(f_1^4 + 2f_1^2 f_2^1 + 3f_2^2 + 2f_4^1) \qquad (32)$$

Substitution of the expression

$$f_k^{\ell} = (1 + x^k + y^k)^{\ell} \qquad (33)$$

into Equation (32). The expression (33) in this form is needed to produce the counting polynomial when k atoms of hydrogen of cyclobutadiene are substituted by the monovalent radical R_1 and k hydrogen atoms of cyclobutadiene are simultaneously substituted by the different monovalent radical R_2.

Table 6
THE CYCLE INDEXES OF SEVERAL BENZENOID HYDROCARBONS

Benzenoid hydrocarbon	Symmetry group	Cycle index
Benzene	D_{6h}	$(1/12) \, (f_1^6 + 4f_2^3 + 3f_1^2 f_2^2 + 2f_3^2 + 2f_6^1)$
Naphthalane	D_{2h}	$(1/4) \, (f_1^8 + 3f_2^4)$
Anthracene	D_{2h}	$(1/4) \, (f_1^{10} + f_1^2 f_2^4 + 2f_2^5)$
Phenanthrene	C_{2v}	$(1/2) \, (f_1^{10} + f_2^5)$
Tetracene	D_{2h}	$(1/4) \, (f_1^{12} + 3f_2^6)$
Triphenylene	D_{3h}	$(1/6) \, (f_1^{12} + 3f_2^6 + 2f_3^4)$

Table 7
THE COUNTING POLYNOMIALS FOR SEVERAL BENZENOID HYDROCARBONS

Benzenoid hydrocarbon	Maximal number of substituents	The counting polynomial
Benzene	6	$1 + x + 3x^2 + 3x^3 + 3x^4 + x^5 + x^6$
Naphthalene	8	$1 + 2x + 10x^2 + 14x^3 + 22x^4 + 10x^6 + 2x^7 + x^8$
Anthracene	10	$1 + 3x + 15x^2 + 32x^3 + 60x^4 + 66x^5 + 60x^6 + 32x^7 + 15x^8 + 3x^9 + x^{10}$
Phenanthrene	10	$1 + 5x + 25x^2 + 60x^3 + 110x^4 + 126x^5 + 110x^6 + 60x^7 + 25x^8 + 5x^9 + x^{10}$
Tetracene	12	$1 + 3x + 21x^2 + 55x^3 + 135x^4 + 198x^5 + 236x^6 + 198x^7 + 135x^8 + 55x^9 + 21x^{10} + 3x^{11} + x^{12}$
Triphenylene	12	$1 + 2x + 14x^2 + 38x^3 + 90x^4 + 132x^5 + 166x^6 + 132x^7 + 90x^8 + 38x^9 + 14x^{10} + 2x^{11} + x^{12}$

The counting polynomial

$$Z(\text{cyclobutadiene}) = 1 + x^4 + x^3 y + 2x^2 y^2 + xy^3 + y^4 \qquad (34)$$

Structures of all positional isomers of cyclobutadiene with two different monovalent substituents R_1 and R_2

The Pólya's theorem was generalized by several people,[54-60] who, thus, have extended the power and the spectrum of uses of the theorem.

Since its introduction of the Pólya's theorem has been, and is, widely used for enumeration of various isomeric, anorganic, and organic, structures. Thus, for example, it was employed for enumeration of isotopic isomers,[48,61] cyclic molecules,[48,62] benzenoid hydrocarbons (polyhexes, arenes),[63] porphyrins,[64] chiral and achiral alkanes,[65] ferrocenes,[66] cluster compounds,[67] various inorganic structures,[68] etc. Certainly the Pólya's theorem is the most important enumeration technique available, although in the recent years the computer oriented methods are used in many instances because of the advantages that the high-speed computers brought to this area of research:[22,69-73] high accuracy of numerical work and possibility to display the generated structures.

REFERENCES

1. **Biggs, N. L., Lloyd, E. K., and Wilson, R. J.,** *Graph Theory 1736—1936,* Clarendon, Oxford, 1976, chap. 4; see also **Rouvray, D. H.,** *Chem. Soc. Revs.,* 3, 355, 1974.
2. **Liu, C. L.,** *Introduction to Combinatorial Mathematics,* McGraw-Hill, New York, 1968.
3. **Schonland, D. S.,** *Molecular Symmetry,* Van Nostrand, London, 1965.
4. **Streitwieser, A., Jr., and Heathcock, C. H.,** *Introduction to Organic Chemistry,* Macmillan, New York, 1976, 62.
5. **Berzelius, J. J.,** *Ann. Phys. Chem.,* 19 (der ganzen Folge 95), 305, 1830.
6. **Senior, J. K.,** *J. Chem. Phys.,* 19, 865, 1951.
7. **Pimentel, G. C. and Spratley, R. C.,** *Understanding Chemistry,* Holden-Day, San Francisco, 1971, 733.
8. **Butlerov, A. M.,** *Z. Chem.,* 5, 298, 1862.
9. **Cayley, A.,** *Phil. Mag.,* 13, 172, 1857.
10. **Cayley, A.,** *Phil. Mag.,* 47, 444, 1874.
11. **Henze, H. R. and Blair, C. M.,** *J. Am. Chem. Soc.,* 53, 3042, 1931.
12. **Henze, H. R. and Blair, C. M.,** *J. Am. Chem. Soc.,* 53, 3077, 1931.
13. **Pólya, G.,** *Acta Math.,* 68, 145, 1937.
14. **Domb, C.,** in *Phase Transitions and Critical Phenomena,* Vol. 3, Domb, C. and Green, M. S., Eds., Academic Press, London, 1974, 1.
15. **Kirchhoff, G.,** *Ann. Phys. Chem.,* 72, 497, 1847.
16. **Cayley, A.,** *Phil. Mag.,* 18, 374, 1859.
17. **Jordan, C.,** *Journal Reine Angew. Math.,* 70, 185, 1869.
18. **Sylvester, J. J.,** *Proc. R. Soc. Inst. G. B.,* 7, 179, 1873—1875.
19. **Cayley, A.,** *Rep. Br. Assoc. Adv. Sci.,* 45, 257, 1875.
20. **Cayley, A.,** *Am. J. Math.,* 4, 266, 1881.
21. **Otter, R.,** *Ann. Math.,* 49, 583, 1948.
22. **Knop, J. V., Müller, W. R., Jeričević, Ž., and Trinajstić, N.,** *J. Chem. Inf. Comput. Sci.,* 21, 91, 1981.
23. **Cayley, A.,** *Ber. Dtsch. Chem. Ges.,* 8, 1056, 1875.
24. **Cayley, A.,** *Philos. Mag.,* 3, 34, 1877.
25. **Schiff, H.,** *Ber. Dtsch. Chem. Ges.,* 8, 1542, 1875.
26. **Herrmann, F.,** *Ber. Dtsch. Chem. Ges.,* 13, 792, 1880.
27. **Lozanitsch, S. M.,** *Ber. Dtsch. Chem. Ges.,* 30, 1917, 1898.
28. **Herrmann, F.,** *Ber. Dtsch. Chem. Ges.,* 30, 2423, 1898.
29. **Lozanitsch, S. M.,** *Ber. Dtsch. Chem. Ges.,* 30, 3059, 1898.
30. **Herrmann, F.,** *Ber. Dtsch. Chem. Ges.,* 31, 91, 1898.
31. **Tiemann, A.,** *Ber. Dtsch. Chem. Ges.,* 26, 1595, 1893.
32. **Delannoy, M.,** *Bull. Soc. Chim. Fr.,* 239, 1894.
33. **Goldberg, A.,** *Chem.-Ztg.,* 22, 395, 1898.
34. **Blair, C. M. and Henze, H. R.,** *J. Am. Chem. Soc.,* 54, 1098, 1932.
35. **Blair, C. M.,** Ph.D. thesis, University of Texas at Austin, Austin, 1933.
36. **Perry, D.,** *J. Am. Chem. Soc.,* 54, 2918, 1932.
37. **Blair, C. M. and Henze, H. R.,** *J. Am. Chem. Soc.,* 54, 1538, 1932.
38. **Henze, H. R. and Blair, C. M.,** *J. Am. Chem. Soc.,* 55, 680, 1933.
39. **Coffman, D. D., Blair, C. M., and Henze, H. R.,** *J. Am. Chem. Soc.,* 55, 252, 1933.
40. **Coffman, D. D.,** *J. Am. Chem. Soc.,* 55, 695, 1933.

41. **Henze, H. R. and Blair, C. M.,** *J. Am. Chem. Soc.,* 56, 157, 1934.
42. **Kornilov, M. Y.,** *Zh. Strukt. Khim.,* 8, 373, 1967.
43. **Davis, C. C., Cross, K., and Ebel, M.,** *J. Chem. Educ.,* 48, 675, 1971.
44. **Polansky, O. E.,** *Math. Chem. (Mülheim/Ruhr),* 1, 11, 1975.
45. **Balaban, A. T.,** *Math. Chem. (Mülheim/Ruhr),* 1, 33, 1975.
46. **Harary, F., Palmer, E. M., Robinson, R. W., and Read, R. C.,** in *Chemical Applications of Graph Theory,* Balaban, A. T., Ed., Academic Press, London, 1976, 11.
47. **Read, R. C.,** in *Chemical Applications of Graph Theory,* Balaban, A. T., Ed., Academic Press, London, 1976, 25.
48. **Balaban, A. T.,** in *Chemical Applications of Graph Theory,* Balaban, A. T., Ed., Academic Press, London, 1976, 63.
49. **Rouvray, D. H. and Balaban, A. T.,** in *Applications of Graph Theory,* Wilson, R. J. and Beineke, L. W., Academic Press, London, 1979, 177.
50. **Slanina, Z.,** in *Advances in Quantum Chemistry,* Vol. 13, Löwdin, P.-O., Ed., Academic Press, London, 1981, 89; see also **Dimitriev, I. S.,** *Molecules Without Chemical Bonds,* Mir, Moscow, 1981, 138.
51. **Redfield, J. H.,** *Am. J. Math.,* 49, 433, 1927; see also **Davidson, R. A.,** *J. Am. Chem. Soc.,* 103, 312, 1981.
52. **Lunn, A. C. and Senior, J. K.,** *J. Phys. Chem.,* 33, 1027, 1929.
53. **Cotton, F. A.,** *Chemical Applications of Group Theory,* John Wiley & Sons, New York, 1964, third printing.
54. **Riordan, J.,** *J. Soc. Ind. Appl. Math.,* 5, 225, 1957.
55. **Kennedy, B. A., McQuarrie, C. H., and Brubaker, C. H.,** *Inorg. Chem.,* 3, 265, 1964.
56. **Harary, F. and Palmer, E. M.,** *J. Comb. Theory,* 1, 157, 1966.
57. **Williamson, S. G.,** *J. Comb. Theory,* 8, 162, 1970.
58. **McDaniel, D. H.,** *Inorg. Chem.,* 11, 2678, 1972.
59. **Palmer, E. D.,** in *New Directions in the Theory of Graphs,* Harary, F., Ed., Academic Press, New York, 1973, 187.
60. **Harary, F. and Palmer, E. M.,** *Graphical Enumeration,* Academic Press, New York, 1973, 33.
61. **Balaban, A. T.,** *J. Labelled Comp.,* 6, 211, 1970.
62. **Balaban, A. T. and Harary, F.,** *Rev. Roum. Chim.,* 12, 1511, 1967.
63. **Rouvray, D. H.,** *J. S. Afr. Chem. Inst.,* 26, 141, 1973; 27, 20, 1974.
64. **Balaban, A. T.,** *Rev. Roum. Chim.,* 20, 227, 1975.
65. **Robinson, R. W., Harary, F., and Balaban, A. T.,** *Tetrahedron,* 32, 355, 1976.
66. **Rinehart, K. L. and Motz, K. L.,** *Chem. Ind.,* 1150, 1957.
67. **King, R. B.,** *J. Am. Chem. Soc.,* 94, 95, 1972.
68. **Krivoshei, I. V.,** *Zh. Strukt. Khim.,* 4, 757, 1963; 6, 322, 1965; 7, 430, 1966; 7, 638, 1966; 8, 321, 1967.
69. **Lederberg, J., Sutherland G. L., Buchanan, B. G., Feigenbaum, E. A., Robertson, A. V., Duffield, A. M., and Djerassi, C.,** *J. Am. Chem. Soc.,* 91, 2973, 1969.
70. **Masinter, L. M., Sridharan, N. S., Lederberg, J., and Smith, D. H.,** *J. Am. Chem. Soc.,* 96, 7702, 1974.
71. **Kudo, Y. and Sasaki, S.-I.,** *J. Chem. Inf. Comp. Sci.,* 16, 43, 1976.
72. **Kudo, Y., Hirota, Y., Aoki, S., Takada, Y., Taji, T., Fujioka, I., Higashino, K., Fujishima, H., and Sasaki, S.-I.,** *J. Chem. Inf. Comp. Sci.,* 16, 50, 1976.
73. **Knop, J. V., Szymanski, K., Jeričević, Ž., and Trinajstić, N.,** *J. Comp. Chem.,* in press; **Knop, J. V., Szymanski, K., Jeričević, Ž., and Trinajstić, N.,** *Int. J. Quantum Chem.,* in press; **Trinajstić, N., Jeričević, Ž., Knop, J. V., Müller, W. R., and Szymanski, K.,** *Pure Appl. Chem.,* in press.

INDEX